W9-BVL-737

...ARY

...vesti...g

...orm th...

...o...

...good.

...A 94...

...555...

www...

Burro Genius

OTHER WORKS BY **Victor Villaseñor**

FICTION
Macho!

NONFICTION
Jury: People vs. Juan Corona
Rain of Gold
Walking Stars
Snow Goose: Global Thanksgiving
Wild Steps of Heaven
Thirteen Senses

SCREENPLAYS
Ballad of Gregorio Cortez

VICTOR VILLASEÑOR

Burro Genius

a memoir

rayo *An Imprint of* HarperCollins*Publishers*

BURRO GENIUS. Copyright © 2004 by Victor Villaseñor. All rights reserved. Printed in the United States of America. No part of this book may be used or reproduced in any manner whatsoever without written permission except in the case of brief quotations embodied in critical articles and reviews. For information, address HarperCollins Publishers Inc., 10 East 53rd Street, New York, NY 10022.

HarperCollins books may be purchased for educational, business, or sales promotional use. For information, please write: Special Markets Department, HarperCollins Publishers Inc., 10 East 53rd Street, New York, NY 10022.

Grateful acknowledgment is made for permission to reprint from "(Ghost) Riders In The Sky (A Cowboy Legend)," from *RIDERS IN THE SKY,* by Stan Jones; © 1949 (Renewed) EDWIN H. MORRIS & COMPANY, A Division of MPL Communications, Inc. All rights reserved.

FIRST EDITION

Book Design by Shubhani Sarkar

Printed on acid-free paper

Library of Congress Cataloging-in-Publication Data
Villaseñor, Victor.
    Burro genius : a memoir / Victor Villaseñor.—1st ed.
        p. cm.
    ISBN 0-06-052612-2
    1. Villaseñor, Victor.   2. Authors, American—20th century—Biography.
3. Mexican Americans—California—Biography.   4. Mexican American
authors—Biography.   5. California—Biography.   I. Title.
PS3572.I384Z464   2004
813'.54—dc22
[B]   2003069346

04   05   06   07   08   DIX/QWF   10   9   8   7   6   5   4   3   2   1

This book is dedicated to Ramón and the guys from *Pozole* Town, Oceanside, and to all boys and girls the whole world over who went to school with laughter in their eyes, warmth in their hearts, kindness in their souls, and then were systematically "broken" of their spirit—their genius, but here and there, some were able to refind that spirit because of an angel-teacher who helped give them back their wings. To these brave souls, both students and teachers, I dedicate this book.

The Kingdom of God is within us all.

Jesus Christ

What lies before us and what lies
behind us are small matters compared
to what lies within us. And when we
bring what is within out into the world,
miracles happen.

Ralph Waldo Emerson

Burro: a small horse, a donkey, biblically called an "ass," one of the older and most widely used animals for work and burden throughout much of the world. An animal used to breed with horses to get a mule, a larger, stronger beast of burden than a burro.

Genius: guardian deity, or spirit of a person; spirit, natural ability. According to ancient Roman beliefs, a guardian spirit assigned to a person at birth; tutelary deity, hence the guardian spirit of any person, place, etc. A person having great mental capacity or an inventive ability; especially great and original creative ability in art, science, etc.

Preface

In the spring of 1962, I started writing this book. I was twenty-two years old. I'd been writing short stories for two years, so I figured that I was ready for the big book. I went through six different drafts of this book that first year, and I mailed out each draft to a publisher, but I only got rejected. The book became an obsession. With each rejection, I became more excited, because I'd now see how to rewrite the book and maybe, just maybe make it a little bit better.

I began to get up at two or three in the morning and work for twelve to fifteen hours a day. I'd get so emotionally drained by the writing, that I'd feel sick at the end of the day. My family became worried about me and they invited a friend, who was a writer in Los Angeles, to see me. He told me that he'd heard how serious I was about my writing, so he was willing to take a little time off of his busy schedule to glance at my work. I gave him the latest version of my manuscript. He took it home and came back to see me the following week. His face was long. He told me that he was sorry to say this but, as a family friend, he had the obligation to be truthful, so he'd tell me straight out that I had no talent. The book was terrible. And also, I was trying to write way beyond my mental capabilities.

He then explained to me that the world of writing was very competitive and so it would be wrong for him to give me any encouragement. He told me that the best thing for me to do was quit writing, not waste any more of my young life, and take advantage of my father's business enterprises. I thanked him. I could see that he'd hated to say what he'd said to me. Then he said the strangest thing. He looked at me in the eyes and said, "You're not going to pay attention to anything I've just said, are you?"

"No, I'm not," I said.

He shook his head. I could see that he was really worried. A few months later, I went into the Army. I took the last version of my manuscript with me. I tried to work on it while doing my military service. Getting back from overseas, I leaped back into the book again with what I'd learned overseas, writing at an insane pace for a few more years. Finally, I quit the book, took the tools that I'd developed as a writer, and wrote three other books, and also sent these to publishers. I only received more rejections. Then I was almost thirty years old when I wrote *Macho!,* and finally got published after 265 rejections. Immediately, I returned to this book you're reading, thinking I could now pull it off, but I was wrong.

Over the next few decades, I sold other books, but I'd always return to this book. I accumulated more that a hundred different drafts. I got married, became a national best-seller author, had two sons, but still no matter how much I'd try to pull this book off, I just kept missing the mark, even though some publishers were saying that some parts of the book were beautiful. Then my father passed over to the Other Side, and my mother passed, too, and somehow, with both of them now in the Spirit World with my brother Joseph, the Voice within me grew stronger, crystal clear, and I was now able to complete this book.

Thank you. *Gracias.* It took me forty years of wandering through the jungles of my mind, heart, and soul to finally realize this book. Enjoy. From *mi familia* to your *familia.*

BOOK one

CHAPTER **one**

I'd been writing for thirteen years, received over 260 rejections, and had just gotten—thank God—my first book published! The year was 1973. I was thirty-three years old, in Long Beach, California, at a CATE conference, meaning California Association of Teachers of English. I was in the back room along with five other writers. All of the other authors had previous works published. We were waiting for the main speaker to show up. This writer wasn't only published, like the rest of us; no, he'd had a best seller, was a nationally recognized speaker, and was going to show up any minute and give the keynote address to the luncheon of the whole CATE convention.

Karen, our publisher's publicist, was nervous as hell, pacing the room and trying to figure out what to do. The national best-selling author should've arrived at least thirty minutes ago. He was supposed to have flown in from the East Coast the night before on the red-eye.

Myself, I was pretty nervous, too. I'd never been in a room with so many writers before. In fact, I'd never even met a published author until about six months back, and that was when I'd been in the Los Angeles office of my New York publisher and I'd finally found out that yes, yes, yes, I was really going to be published! I immediately called my mother and father, screaming to the high heavens—I'd been so excited. Bantam from New York was going to publish my book *Macho!*

The room we were in was small, but felt much larger because of all the excitement. I had no idea what was expected of me, so I stood in a corner by myself, playing it safe and just watching everything. Hell, the only reason I was even here was because our publicist Karen Black— who was actually white—had called me up out of the blue yesterday afternoon, I guess, as an afterthought, and said, "Don't you live just south of Long Beach?"

"Yes, I do," I'd said.

"Good. I hope you're not too busy or will take offense that I'm calling you so late, but you see, we're going to have several of our authors giving workshops at a CATE conference in Long Beach this weekend, so why don't you drive up the coast and join us?"

"Cat? What's that?" I'd asked.

"No. CATE, California Association of Teachers of English. They buy a lot of books. This conference is very important for us, and could be for you, too."

"Oh, I see. Yeah, sure, I'll come," I said, taking a deep breath. "Will I be attending one of the workshops?"

"We thought you might present a workshop."

"Me?"

"Yes, of course. You are a published author."

My heart began pounding. "What would I give a workshop on to English teachers?"

"On your experiences in writing. On that special English teacher who inspired you to become an author," she said full of honey. " 'Bye now. We'll see you there. Don't worry. You have a creative mind. You'll come up with something."

She gave me the address, and then this morning, I drove in my white van up from Oceanside, where I still lived on the ranch on which I was raised, to Long Beach. I'd never heard of CATE in all my life, much less did I know what it meant to "present" a workshop. All I knew was that I'd flunked the third grade twice because I couldn't learn to read, had a terrible time all through grammar school and high school. Then after ten years of writing, I was finally able to sell my first book to a leading mass-market paperback publisher in New York.

And now, standing in a corner, I felt pretty green. After all, these other writers in the room had been published before and they were talking to one another like they were all best friends, swapping publishing stories, laughing happily, eating cookies and drinking coffee.

I was drinking water. One sip of coffee would have shot me through the roof. Listening to the conversation around the snack table, I was beginning to understand that these other writers had not only already had several books published, but that most of their books had first come out in hardback, then had come out in mass-market paperback.

I was quickly learning that it was not very prestigious for me to have first been published in paperback. Because paperback books didn't get reviewed, and reviews were what got an author attention, respect, and sold books. Hell, I was still so wet behind the ears that I hadn't even realized what a review was until a few weeks back. So I said nothing and just kept listening closely, trying to learn all I could without showing my ignorance. Also, I could now see that these other writers were dressed more like city people. I guess that it had been a mistake for me to come in Levi's, cowboy boots, a big belt buckle, a Western shirt, and my old blue blazer.

Behind the closed doors of the next room, we could hear the low, rumbling noise of all the people at the conference eating lunch. I figured that it had to be a good-size crowd of people by the sound of the ruckus of plates and conversation. Our publisher's publicist was now chain-smoking as she paced the room. Checking her watch for the umpteenth time, Karen now sent her assistant, Sandy, to check for any messages at the lobby, then told her to also go out to the parking lot and glance around. Boy, it was all like a movie. Here I was in the back room with a bunch of real writers, and any second now a nationally recognized author was going to come rushing down the hallway and lead us through the two closed doors where a whole convention of teachers was waiting to meet us.

My heart was pounding a good million miles an hour with all the excitement. After all these years, I was really a published author, standing there eating carrot sticks right along with other published writers. Before this, all the writers I'd ever seen were on the back covers of books and on posters up on the walls in libraries and bookstores. I had to keep pinching myself to make sure that all this was really true. Every day, for over thirteen years, I'd been dreaming of something like this.

Suddenly Sandy came rushing back into the room, handing Karen a note. Our publicist glanced at the note, and it looked like she was going to scream, but she didn't, and gave a little curse instead.

"Damnit!" she said. "He didn't take the red-eye! His plane has just landed! His limo driver says that he can't have him here for at least another forty minutes. We can't keep this convention waiting any longer!"

Wow, this was all becoming more like a fast-paced movie by the moment, I thought. But then, the next thing I knew, Karen turned her attention to us, the writers, who were across the small beige-green room by the coffee table laid out with snacks.

"Have any of you ever been a keynote speaker?" she asked.

None of us answered. We just glanced at one another.

Seeing our reaction, Karen crossed the room in large, confident steps. I could feel her determination. She was going to get something out of one of us, and immediately, too.

"Look," she said to us in a calm and yet forceful voice, "behind those two green doors we have a convention room full of English teachers. They've been patiently waiting for over thirty minutes. I need to give them a speaker in the next five minutes or less. Can anyone of you handle this situation?"

I glanced around at my fellow authors, and I couldn't believe it. Not one of these writers, who'd had hardcover books published and knew a hell of a lot more about what was going on than me, was coming forward. I took a deep breath, straightened up, picked up my Western hat off a chair, and stepped forward.

"I can do it!" I said in a loud, clear voice.

She looked at me, saw my hat in my hand—which I was very proud of because my dad had given it to me—glanced at my shirt and jeans and down to my boots.

"Which title is yours?" she asked.

"*Macho!*" I said.

"It's about Chicanos, right?"

"No, not really. It's about a young Mexican boy coming across the border without papers to the United—"

"Chicanos, Mexicans, they're all just about the same, right?"

"Well, yeah, in a way, except they're totally different, because one is born in the United States and the other is born in—"

"Have you ever spoken publicly before?"

"Well, no, not really, but I know I can do it just like, well, I knew that I'd someday get published."

"And how long did that take you?"

I didn't appreciate the way she was looking at me. "Not very long," I said, lying.

She turned away from me. "Haven't any one of you other writers ever been a keynote speaker?" she asked.

The most smartly dressed writer among us, a woman, who was probably—I guess—in her late thirties or early forties, spoke up. "I have," she said. "Several times. But I was also given notice beforehand, so I'd have time to prepare."

"Don't you have something prepared for your workshop?"

"Certainly, of course. But as I'm sure you realize, a workshop and a keynote address are very different," she said, raising her left eyebrow in one of the most dignified arches that I'd ever seen.

Karen turned back to me. "You do realize," she said, "that public speaking is very different from writing. Few authors are good public speakers. One is done in private and the other—maybe we'd better just wait," she said, turning to her assistant Sandy.

I was beginning to like Karen less and less. Yeah, sure, she was good-looking and smartly dressed, too, and had probably been the top student in her class all through grammar school, high school, and college, but she just didn't seem very open-minded or flexible. I liked her assistant Sandy a lot better. She seemed softer, less judgmental, and more open. I guessed that Karen was single and in her mid-thirties. Sandy, I bet, had a boyfriend and was in her late twenties and had had more than just straight *A*'s going for her in school. Myself, I'd mostly been a C student, had wrestled, worked on the ranch, and now had a girlfriend to whom I'd recently proposed, but she'd turned me down. Also, I had such a baby face that, until about two years ago, I was still ID'd regularly in most bars.

Sandy quickly whispered something to Karen. "Okay! Okay!" said our publicist. "Great idea! Good thinking! You," she said, turning back to me, "come with me, but that hat has to go! And could we get him a different shirt? These are educated people."

I closed my eyes so I could concentrate on the fact that, I guess, it had just been decided that I'd be the one to give the keynote address, so what I was wearing wasn't really the main issue.

I followed Karen and Sandy down the hallway. But then I suddenly realized that I had to pee. This had also happened to me in high school when I'd been a wrestler. I'd get real nervous just before a match and have this terrible feeling that I needed to pee. Then I'd go to the bathroom and I couldn't pee for the life of me. But I was afraid to say anything about this to Karen and Sandy, so I just kept following the two women down the long, narrow hallway.

Sandy kept whispering in Karen's ear, and glancing back at me. I liked Sandy. She didn't just look at me like I was a head of lettuce that didn't fit in with the other heads of lettuce. I was able to catch the tail end of their conversation. Sandy was reminding Karen that this was the West Coast and not New York, so a Mexican-American writer might just fit in very well with this group of teachers, especially with all the daily news going on about Cesar Chavez and his fieldworkers.

"All right," said Karen, "maybe it will work. But what happened to our backup speaker? You know that we always have to have a backup speaker when we do these big events."

Sandy glanced at me. "He's in the bar," she said to Karen. "I just saw him. I don't think we can go with him."

"That's all we need, another drunk author. All right, then we'll go with this writer. But do we have anything on him? And who will introduce him?"

"I can do that," said Sandy, smiling at me. "I just read his bio. It has legs."

"Really?" said Karen, sounding impressed.

"Yes."

"All right, then he's our man of the hour." Then I couldn't believe it, Karen now turned to me with a big smile—as if I'd really been her first choice all along and said, "How are you feeling? Is there anything we can do for you?"

I felt like saying, "No, thanks, ma'dam. You've already kicked the crap out of me enough." But I didn't. Instead I just said, "No, I just need a bathroom, then I'm ready to go."

Her face twisted. "Are you getting sick?"

I had to take a big breath. "No," I said. "I just need to pee."

"Oh, good! That's wonderful! You take him to the bathroom, Sandy, and I'll go tell them that we'll be ready to go on in two minutes. Is that long enough for you to, well, get ready?" she asked me.

I nodded. "Yeah, sure."

"Good, then hurry. This is an important keynote for us. A lot of teachers and entire school districts are going to know about you and your book in just a few minutes. Your English, I mean, you do speak proper English, I take it."

I decided to not tell her about my flunking the third grade twice. I also decided to not inform her that I was never able to get into regular lit classes in college because I could never pass remedial English. I turned on my heels. I'd just about taken all I could from her.

"Please don't take what Karen says too seriously," said Sandy to me as we quickly went down the hallway, by the lobby to the restrooms. "She's really a very fine person once you get to know her. She's just under a lot of pressure. Being a publicist can be hell at times like this."

As we flew pass the bar, I recognized the well-known science fiction writer at the bar, drinking and tossing down peanuts. At the door of the men's room, I abruptly stopped. Sandy was staying so close to me that I thought she might just come in with me.

"I'll wait for you right out here," she said. "And could you please give me your hat so you can comb your hair?"

I didn't want to, but I handed her my hat.

"You do have a comb, don't you?"

"Yes, I do," I said, feeling like I suddenly had a new mother.

"I'm sorry that we're being so pushy," she said, "but as you might guess, it can sometimes be very difficult handling writers."

I nodded and went inside. Immediately, I walked up to a stall and unbuttoned my Levi's. I always wore the old, original button-fly Levi's because of the short crotch. I have long legs for a man my height, but a short torso, so the newer Levi's with a zipper ride too high up on my waist.

I couldn't pee, no matter how much I tried. And I really needed to go, too. Finally, I walked across the huge bathroom. I was all alone. I washed my hands in the sink, then washed my face with cold water. Then I remembered how this girl that I'd met up in San Francisco right after I'd gotten out of the Army would always turn on the water in the sink when she went to pee. I'd thought that she did it because she didn't want anyone to hear her waterfall echoing in the toilet, but she'd explained to me that the sound of running water actually relaxed her so she could pee more easily. I decided to give it a try.

I dried my hands and face, left the water running in the sink, recrossed the bathroom, and put myself in front of a stall once again. Closing my eyes and breathing real slowly, I was finally able to start peeing. And man, it just kept coming and coming, and it was still coming fast when the knocking started on the door. But I just ignored the knocking and kept on peeing. I mean, it was like I was never going to stop.

Then, I couldn't believe it, someone opened the door. "Are you okay?" said a woman's voice. It was Karen. "Everyone is waiting."

"Yes," I said, feeling irritated as hell.

"It's been two minutes, you know."

"Please," I said. "I'm busy! I'll be right out!"

"You're not sick, are you?"

"No, I'm fine. Just shut the door, please."

She closed the door and I finished peeing, buttoned up, went back to the sink, washed my face with cold water again, then turned off the water. I wondered if she'd thought that the sound of the water from the faucet had been me peeing. I laughed. That was really funny. I took another couple of paper towels, dried my hands and face, then glanced into the mirror, saw my wide, high-cheek-boned face, my straight black hair, took two or three deep breaths just as I'd always done in high school before a wrestling match, then said in Spanish—not ever in English—"*Papito Dios,*" meaning little Daddy God, "please, *Papito,* You got to stick close to me right now. I need Your help, no kidding. *Gracias.*"

Saying this, I quickly made the sign of the cross over myself, felt better, and knew that I was now as ready as I'd ever be, *con el favor de Dios.* Hell, my freshman year in high school I'd made the varsity team, won nine out of twelve matches, and I'd been wrestling against juniors and seniors, guys two and three years older than me. I felt if I could do that, I could do this. No *problema.* I turned and walked out of the bathroom.

Both Karen and Sandy were waiting for me right outside of the door. Karen got hold of my arm. She was really strong. Quickly she started walking me as fast as she could back across the lobby, then up the hallway towards the rear entrance of the convention center. I kept trying to get my hat back from Sandy, but Karen finally said, "Absolutely not! We'll keep it for you until after your talk."

But I always wore hats. I'd learned this as a little kid from watching Beeny and Cecil, the sea serpent on kid's TV. Beeny had always put on his thinking hat before he did anything important. And my dad, who'd sometimes taken the time to watch the little kids' TV shows with me, had told me that he entirely agreed with Beeny 110 percent, because he, too, always wore a hat when he played poker or had any important thinking to do.

"Look, I want my hat," I said to Karen as we came to the end of the hallway.

"No," said Karen, taking my hat from Sandy and holding it be-hind herself. "I will not have one of our authors going out there with a ridiculous-looking old hat. It's bad enough that you're wearing that loud Western shirt and big belt buckle. Don't you get it, you're a pub-lished author now. You're in the majors."

I took a big breath. That did it. She'd just pushed one too many of my buttons. Women got to wear colorful clothes and even padded bras in the majors, didn't they? So why couldn't a guy wear a bright turquoise shirt and a big belt buckle? Hell, her bra was probably even padded. Sandy opened the door for us, and suddenly here I was along with Karen and the five other authors in a gigantic room that stretched out forever in all directions. The whole place was packed full of people sitting at large round tables, about ten to a table. It looked like acres and acres of people, and you could see that they'd already eaten, were into their dessert and coffee, and were looking pretty antsy.

I froze. This was one hell of a lot bigger affair than I'd ever imag-ined. Sandy immediately went to the microphone. She had a copy of my book *Macho!* in hand. Glancing at the back cover, she began to read, introducing me to the audience. I glanced around. I really needed my dad's hat right now, at least in my hands. I glanced at Karen. She was smiling radiantly at the audience and she now had my hat behind her shirt, like she was trying to hide it from me. I took a big breath. She'd been treating me like a little kid ever since they'd walked me to the bathroom.

Sandy was almost done with my introduction. She was now telling the crowd that I was one of their newly published authors. "He's writ-ten an excellent book called *Macho!* but it isn't really about *machismo*. It tells the story of a young Tarascan Indian boy who journeys from Michoacán, Mexico, to the United States and toils in the fields with Cesar Chavez." Everyone started clapping, including Karen, and in that split second, with her holding my hat in front of her body as she clapped, I leaped forward and snatched it from her, then turned on my heels and walked up to the mike before she could say or do anything.

I was terror-stricken, but felt better now that I had my dad's hat with me. This was my padded bra with which to face the world. I put on the old sweat-stained Stetson and instantly felt better. Then I took hold of Sandy, who was handing me the microphone, and I gave her *un abrazo*, meaning a hug, wanting to thank her for her beautiful intro- duction. But I felt her tense her body and start to panic. Then, hearing my words of thanks in her ear, she relaxed and hugged me back. This felt good. I was set. I'd gotten to that calm, safe place inside of me that I'd needed to get to. She turned me around but didn't hand me the mike. Instead she slipped it into the metal gadget that was built into top of the boxlike podium, which was made of beautiful oak. I breathed. She'd been smart to do this. My hands were shaking too much for me to have handled the mike. But I wasn't really very worried. I'd had some of my best wrestling matches when I'd been this wound up.

Sandy left my side and now I was all alone, standing before one of the largest crowds of people that I'd ever seen assembled, except of course, at the Del Mar racetrack. I had no idea what to do, much less where to begin. In the bathroom, I'd thought of maybe just opening up my book and reading to them about the Paricutín Volcano in the state of Michoacán, Mexico, where my novel begins.

But then, I'd also remembered how all my life, I'd had so much trouble reading, especially aloud, that this wouldn't be the best bet for me. Hell, by the time I'd gotten to the fifth grade I was so gun-shy of reading aloud that I'd rather have had the teacher hit me with her ruler than put me through the embarrassment of all the kids finding out what a terrible reader I was.

My head was beginning to get hot. I took off my hat, put it on the podium, and continued to look out at the crowd of mostly all Anglo peo- ple. There were only a few Blacks scattered here and there. Nowhere did I see a wide brown face like mine.

I took another big breath and decided to just talk, to simply tell them how I'd researched this book by interviewing some of the guys on our ranch, and then how I'd combined the story of one main guy with two other of our *vaqueros*. Then I'd explain that I'd found out that inter- viewing people didn't quite do it, so I'd then had to actually cross the border myself, illegally, at Mexicali, and work in the field picking melon from the border up through Bakersfield and then Fireball, the world capital *del melon*.

Then I don't know what exactly happened to me, but as I looked out at this sea of faces, and I realized that they were all . . . English teachers, I suddenly felt my heart EXPLODE! But not with fear. No, with white-hot rage, and I now knew exactly what it was that I really wanted to say.

I took in another deep breath. Some of these teachers were now getting up to leave, but this didn't frighten me. Just as it hadn't frightened me to go into wrestling matches where I had seen that my opponent didn't have much respect for me. In some of those matches I'd gone in with lightning speed, taken the older, more experienced guy by complete surprise, and pinned him within seconds.

"EXCUSE ME!" I shouted, not realizing the microphone would amplify my voice into a thunderous sound. "BUT I UNDERSTAND that all of you here are English teachers!" My booming voice stopped everyone in their tracks. I glanced at Karen and Sandy, who were over to the right of me some twenty feet away. It looked like Karen was just about ready to shit a brick.

I closed my eyes, got my publicists and everything else out of my head, and put myself astride a thousand-pound horse *a lo charro chingón!* Ten feet tall! Faster and stronger than any human in all the world! Opening my eyes, I grabbed hold of the podium before me like I'd do to a calf in calf roping. "Once, I had an English teacher!" I said, feeling my heart leaping into my throat. People smiled. Others laughed. And the ones who'd been getting up to leave now sat back down. "And to this day . . . I hope to God . . . with all my heart . . . that that English teacher DIES A PAINFUL DEATH THAT LASTS AT LEAST ONE WEEK! BECAUSE! BECAUSE!" I barked into the microphone as I grabbed the podium with such force, that I shook its heavy oak up and down. "I can forgive bad parents! Because maybe it was an accident! Maybe they didn't even want to be parents! But teachers are no accident! You study to become a teacher! You work at it for years. So I cannot, and will not forgive teachers who are abusive, mean, and torture kids with commas and periods and misspelling, making them feel like they are less than human because they don't or can't seem to get it right.

"But on the other hand, I pray to God, with all my heart and soul, that all good teachers . . . who are patient . . . attentive . . . considerate and kind go to heaven when they die and they are rewarded with vanilla ice cream and apple pie FOR ALL ETERNITY! Because, you see, a bad teacher, like Moses, the abusive English teacher that I had in Carlsbad, kills kids, here in the heart, and not just with their tests, but with their superior attitude and those sly smiles that they like to give to their *A* students, but never to the ones who are also working hard, or maybe even harder, like me, but just couldn't get it!

"I was TORTURED by teachers! You hear me, TORTURED!" I yelled, jerking the whole podium off the floor. "Hell, I flunked the third grade twice because—BECAUSE—" I was crying so hard that I had to wipe the tears out of my eyes with the back of my hand, but this wasn't going to stop me. I was all guts up front now. I was in that smooth-feeling, all-true place that I got into when I'd go to my room and start writing each morning before daybreak . . . with all my heart and soul.

"TRULY!" I shouted. "UNDERSTAND that this, that I'm talking about, is IMPORTANT! I wrote for ten years before I got published! I'd written over six books . . . sixty-five short stories and four plays . . . and received more than two hundred sixty rejections before I got published! And it was hate and rage towards abusive teachers that kept me going year after year . . . with the hope that one day I'd get published, and have a voice, so I could make a difference down here in our hearts and guts," I said, grabbing hold of my own gut, "where we really live, if we're going to live a life worth living! Because, you see, real teaching isn't just about teaching the brain up here," I said, hitting my forehead, "but it's also about inspiring your students to have heart . . . compassion, guts, understanding, and hope!

"My grandmother—God bless her soul—a Yaqui Indian from northern Mexico, was the greatest teacher I'd ever had! And do you know what she taught me, she taught me that each and every day is *un milagro* given to us by God, and that work, that planting corn and squash with our two hands is holy. She taught me all this with kindness and invitation. Not with ridicule and looking down her nose at me and making me feel like less than human when I didn't get it at first."

I was crying so hard that I had to stop talking to catch my breath. Quickly, out of nowhere, Sandy was at my side, patting me on the back and handing me a glass of water. I drank the whole glass down. I suddenly had to pee again, but I figured that I could probably hold it, I hoped.

"Are you okay?" asked Sandy, stroking my arm. "Can you go on?"

I closed my eyes and took a deep breath. "Yes," I said, feeling my heart beat, beat, beating like a mighty drum. "I can go on. I got to!" I added. My grandmother was here at my side. I could feel-see sense her, completely.

"Okay," said Sandy. "You're doing fine. This isn't exactly what they were expecting, but . . . you do have their attention."

I almost laughed. Sandy was right. Looking out at the crowd, I could see that I really did have everyone's attention. In fact, some were at the edge of their seats, ready to take in my every word. But others were shaking their heads, and looking like they wanted to get up and leave. I glanced at Karen, who looked like she'd already shit her brick, and she was now running her index finger across her throat. I guessed that she was telling me that she was going to cut my throat once she got hold of me. I only laughed at her, too. Hell, there was no way on earth that I was now going to be silenced. I'd been holding back all this fire inside me ever since I'd started school.

A whole bunch of people towards the back of the room were now getting up to leave. But these teachers were not going to intimidate me. I wasn't my father's and mother's son for *nada*-nothing. I wasn't my two indigenous grandmothers' grandson for *nada*-nothing either. I came from a long line of people . . . who'd lived through starvation, revolutions, and massacre.

"AND YOU!" I yelled into the microphone. "Over there in the back . . . who are getting up to leave . . . I'M GLAD THAT YOU'RE

LEAVING! Because, obviously, you're those bad teachers that I'm talking about! Hell, if you were good teachers you'd be happy with what I'm saying!" Three fourths of the teachers who'd been getting up to leave suddenly sat back down. I loved it. I was finally kicking ass, after all these years of getting my own ass kicked. "When I started school," I continued, "I spoke no English. Spanish only. And we weren't invited in to learn English in a nice, civilized way. No, we were screamed at. No joke. Yelled at, 'No Spanish, English only,' and then ridiculed, called names, and hit on the head or slapped across the face if we were ever caught speaking Spanish. AND SPANISH WAS ALL I KNEW! And on that first day of kindergarten, I needed to go to the bathroom, but the teacher yelled NO! So I peed all over myself, with pee running down my legs until it formed a puddle by my shoes, and all the kids sitting close to me looked at me in disgust, holding their noses, then at recess they let me have it with ridicule!

"THAT'S NOT RIGHT! Teachers and their arrogant ways pit us kids against each other! A STUDENTS AGAINST ALL OTHER KIDS! B students feeling good that they're not C and D students. CONFRONTATION and WAR is what you teachers start instilling in kids from kindergarten on! 'We need to study history so we don't repeat it' is a BUNCH OF BULL! Don't you get it?" I said, wiping my eyes. "We keep repeating history because that's what you teach—meanness, greed, and every man and woman out for themselves, without mercy!"

Now two whole tables of teachers got up and started to leave.

"SIT DOWN!" I barked. "You're not dismissed! Class is still in session! Hell, I'm just getting started! You came to hear a writer talk, and now—by God—you're going to hear A WRITER SPEAK!"

"I WILL NOT SIT DOWN and listen to any more insults!" yelled one teacher to me from about one third back in the room. "You're just a spoiled, overpaid author and all of us here are underpaid and overworked teachers, every day working above and beyond duty!"

"I got $4,500 for my book that just got published!" I said. "It took me over three years to write it. After my agent's fee, that's less that $1,500 a year. I'm not complaining and saying that I'm underpaid. I'm happy! BIG HAPPY, that I got published! I think it's an honor to be a writer, just as it is an honor to be a teacher, a person who can—"

"You just speak generalities! You're not telling us anything specific!" she yelled, cutting me off.

"You want something specific," I said. "Well, I specifically DON'T LIKE YOU! Because, if you had any heart and didn't feel so guilty for being one of these abusive teachers, you wouldn't take offense at what I'm saying! SO GO! LEAVE! You and all your other offended teachers."

A couple of teachers applauded me, a few tables away from the people who were leaving. But the teacher who spoke, and about a dozen other teachers, left with a big ruckus, taking their book bags, coats, and purses.

Well, about 90 percent of all the other teachers stayed, and when I was through with my talk—about thirty minutes later—a good third of them were also wiping their eyes. When I closed, I got a standing ovation of such volume and magnitude that the walls of the huge room actually vibrated.

Teachers rushed to me. Karen and Sandy had to pull people away from me so that they could take me to our booth down the hall into another huge room so I could sign books. The number of people who got in line to have me sign my book was so long that it dwarfed the lines of all the other writers. And this room was full of other top-name New York publishers with their own writers, too.

Teachers kept asking me if I could come to their schools to give a talk, that I'd been so real, and reality was what their students needed above all else. Others asked if I'd written any books on education. I said, no, not yet, that *Macho!* was my first book.

And Karen, she now couldn't do enough for me. She kept bringing me water, snacks to eat, stroking me on the arm and shoulder, and asking me if I needed anything to give me the strength to keep autographing books. I signed books for well over three hours. And Sandy, when I was done, came up and gave me a big *abrazo*-hug, kissing me on the cheek.

"I just knew you could do it!" she said, beaming. "After reading your bio, I just knew that anyone who'd gone through so much for so long to get published had to have a floodgate holding back all that you needed to say. You did great! In fact, we ran out of your books. And the LA *Times* and the Long Beach newspaper are here to interview you."

"You mean my book is going to get reviewed?" I asked all excitedly.

"Yes, that, too, I'm sure. But right now they're here to do a feature story on you. That's even better than a review. It will reach people beyond those who read the book review section. You're on your way," she added. "Congratulations!" she said, hugging me once again.

Boy-oh-boy, I was flying!

The two interviews were done in the lobby and I had my photo taken too. Then that evening, I was alone in the bar of our hotel having a beer when a man came up to me and said, "Are you the writer who told Gladys Marsh that you specifically didn't like her in front of the whole convention?"

I didn't know what to say. I was exhausted. Really tired. And this man looked like a pretty physical-type guy in his forties who probably had played football and still lifted weights. And hell, he might also just happen to be that woman's best friend or husband. But what the hell, I'd done what I'd done and so there was no backing down now. I got off my bar stool, so he couldn't take a swing at me while I was a sitting duck.

"Yes," I said, taking up ground. "I did say that to a woman teacher, but I don't know if her name is Gladys whatever-you-said or not."

His powerful-looking face suddenly broke into a huge grin and he put out his right hand. "Let me shake your hand and buy you a beer! Hell, I'm part of the teachers' union and that woman is one of the most self-centered, abusive people I've ever had the misfortune of meeting. Always complaining about her paycheck and long hours, but never having any suggestions for helping the kids or improving the system."

We shook hands and he pat me real hard on the back.

"Great job!" he said. "Great job! But tell me, why is it we've never heard of you before? Obviously, you're a gifted speaker and writer, from everything that I've been hearing from everyone who heard you speak and is now reading your book."

And it was true—all afternoon and evening, I'd been seeing people here and there reading my book, and they'd tell me that my book was excellent and that I'd given the best talk that they'd heard in years. My God, what a great feeling this was after so many years of rejection. Man, so much had happened to me since I'd driven up this morning from Oceanside. Suddenly, I felt completely drained.

"I'd like to take a rain check on that beer, if you don't mind," I said. "It's been a big day for me. I need to go to my room."

"Okay," he said, grinning. "Have a good rest. I just wanted to shake the hand of the man who finally spoke up to that ballbuster. Hell, I've known Gladys for fifteen years and no one has ever had the guts to put her in her place."

"Goodnight," I said.

"Goodnight," he answered. "But I have one question to ask you," he added. "Did you really have two hundred and sixty rejections before you were published, or was that a publicity stunt?"

I took a deep breath. "No," I said, "it's no stunt. It's real."

He looked at me in silence for a long time, then said, "But surely there must have been at least one teacher who helped you along the way. It couldn't have all been negative."

I took in another deep breath, thought a moment, then shrugged.

He nodded. "Okay, but think about what I said," he said. "And before your next talk—because I'm sure you are going to become a frequent speaker on the circuit, the way people are responding to you—it would help us immensely if you could also tell us about any good experiences that you might have had.

"Wake-up talks are good, but we also need to know about those teachers who had a constructive influence on you. Believe me, most of my teachers are smart enough and honest enough to admit what's wrong with our system, so we also need to know what works. Congratulations," he said, extending his hand out to me again, "I admire your perseverance, and if you could tell us how you acquired this perseverance, that in itself would help us tremendously."

I nodded. I could see that this man was absolutely right; what I'd given was a wake-up talk and there was a lot more that I still needed to learn—not just about public speaking—but writing, too.

But my God, my wounds were all still so fresh that it was very hard for me to remember any good experiences that I might have ever had at school.

"Thank you," I said to him, and I left the bar and started across the lobby. I was feeling pretty woozy. It was a good thing that I'd accepted the hotel room that Karen had offered me. I could never have made it home. I just didn't know what had come over me when I'd looked out at that sea of English teachers. It was like my heart and soul had leaped forward with so much hate and rage that I'd instantly wanted to kill! To scalp! To massacre! And I'd been fearless, completely trusting my instincts just like I did when I wrote. It was truly a good thing that I'd found writing as my outlet or I was sure I would have become a mass murderer, killing all those heartless, racist teachers who'd beat us Mexican kids down since kindergarten with the complete blessing of our one-sided, dark-ages educational system.

Shit, by the third grade, I'd been so terror-stricken of school, that I'd become a regular bed wetter. I thanked *mi mama* that she'd known how to put me to bed with prayer and song, and that she'd bought a rubber mat to put over my mattress, giving me enough hope and encouragement so I could go on. And I thanked *mi papa* who'd always said to me that we, *los Indios,* the Indians, were like the weeds. That roses you had to water and give fertilizer or they'd die. But weeds, indigenous plants, you gave them *nada*-nothing; hell, you even poisoned them and put concrete over them, and those weeds would still break the concrete, reaching for the sunlight of God. "That's the power of our people," my father would tell me, "we're the weeds, *LAS YERBAS DE TODO EL MUNDO!*"

By the time I'd crossed the lobby and gotten in the elevator, I was so tired I was ready to drop. Hell, I'd never spoken in public before. I hadn't been on the debate team in school. Wow, I'd never had any idea that public speaking could be so tough. In my room, I undressed, cracked open the window, got in bed, and thanked *Papito Dios* for having gifted me another wonderful day in *Paraíso*.

This was also what my old Yaqui Indian grandmother, Doña Guadalupe, had taught me to do each night: to give thanks to God at the end of each day, for every day was, indeed, a paradise given to us by the Almighty. Stardust was what we humans really were, my old grandmother had always explained to me, coming to this *planeta* as Walking Stars just as Jesus had come to do *Papito Dios'* will on this *tierra firma*.

I went to sleep feeling very good because I knew that I'd done my best with all my heart and soul. I didn't feel the least bit bad about how I'd gotten in those teachers' faces and told them what I really thought. After all, hadn't Jesus gotten mad at the money lenders on the temple steps?

But also, I could now understand very clearly that that man from the teachers' union had been right and there was still a lot that I had to learn. Who were those schoolteachers who had helped me? And what was it in my life that had given me the heart . . . the guts . . . to go on and on and never give up, no matter what!

CHAPTER two

Dreaming, I slept in my large, spacious hotel room. Dreaming of all the different waters that had gone slipping, sliding under the bridge on which I'd been living *mi vida*. A bridge bridging my Indian and European roots, a bridge bridging my Mexican and American cultures, a bridge bridging my indigenous beliefs and Catholic–Christian upbringing, a bridge bridging my first few years of life in the *barrio* and then my life on our *rancho grande,* then that whole big world outside of our gates. I slept, feeling so warm and snug underneath the covers of the big hotel bed in Long Beach, the Pacific Ocean slapping the seashore quietly in the near distance as I dreamed . . . remembering back, like in a faraway, foggy dream, that yes, there really had once been a very special teacher who'd helped me along my way. He'd been a substitute teacher and he'd only been with us for two or three days, but my God, he'd touched my very soul in that short time.

I now remembered very clearly that I'd been in the seventh grade. I'd been going to the Army Navy Academy in Carlsbad, California, about four miles south of our *rancho grande*. Originally, I'd started school in Oceanside in an old public grammar school that no longer exists. This school had been located just west of the present Oceanside High School. For five years, I went to public school, but I'd been having such a difficult time that my mother had finally spoken to our priest and it was decided that I'd be sent to Catholic school, thinking that I'd get more personal attention. But it hadn't helped. By the sixth grade, I was still two grade levels behind in reading. This was when it was suggested to my parents by our insurance broker, Jack Thill, that I should be put in our local military school where I'd be taught discipline, and with enough discipline, I would surely learn to read.

On my first day of military school, I was terror-stricken. My older brother Joseph, who'd died when I was nine years old, had gone to this school and my parents told me that Joseph had loved it, but I didn't believe them, because, within the first few minutes of my first day, I was already in deep trouble. Living in Southern California, I'd never worn woolen clothes before. My uniform was made of wool and made me itch so much that I couldn't stand still when they told us to line up, stand at attention, keep our eyes to the front, and stop fidgeting. And my God, I'd thought the nuns were bad where I'd gone to parochial school for three years, but this military school was far worse.

A fellow cadet, not much older than me, put his face in my face and yelled at me to stop fidgeting, that I wasn't a girl, that I was now a cadet of the Army Navy Academy and that I had to stand up tall, pull my shoulders back, keep my eyes to the front, and keep still. I was so terrified that I was ready to pee. I was short so how could I ever stand up tall? And my uniform was making my skin itch so much that I couldn't, for the life of me, stop fidgeting. Still my fellow cadet continued to shout at me right in my face. And his mouth was huge and full of big white teeth.

Within days, I came to realize that this was how cadets got promoted into being corporals and sergeants: by intimidating us other cadets with their in-your-face-shouting. I also came to realize that these cadets, that the school officials put in charge of us new recruits, truly enjoyed their power. In fact, I could now see that some of them had the same little, sneaky smiles on their faces when they shouted at us that I'd seen on the meanest public-school teachers and Catholic nuns.

My mother saved me again, just as she had saved me when she'd bought that rubber mat and put it on my bed so I wouldn't ruin another mattress. She bought some silky material—when I told her about my uniform itching—and she hand-sewed the silky material into the back and front of my shirt and the thighs of my pants. I loved *mi mama,* and she promised to never tell anyone about the baby-blue material that she'd sewed into my uniform, just as she'd promised to never tell any-one about the rubber mat that was on my bed.

I'd been going to the Army Navy Academy for about six months and this morning when I arrived, I figured that it was all going to be the same old routine until we were assembled, lined up, and marched into English class. Our regular English teacher, Captain Moses, wasn't pres-ent. At the front of our room was a man we'd never seen before. He wasn't in uniform. He wore regular civilian clothes and he was short and blond, looked very muscular, and had a huge smile.

"Good morning!" he said to us in a big happy-sounding voice. After we took our seats, he pointed to a large poster that he'd pinned on the wall of someone skiing in the mountains.

"My name is Mr. Swift and I'm your substitute teacher," he said. "Mr. Moses, from what I understand, is home sick. My wife and I are from Colorado . . . where we were . . . ski bums! That's why we are sub-stitute teachers. We moved to Southern California earlier this year to become surf bums, and this is why we will continue to be substitute teachers. I met my wife, Jody, at the University of Boulder, at Boulder, Colorado. She was a track star. I bet that even today, she could give your best sprinters at this school a run for their money in the hundred-yard dash. I've never been able to beat her, and I'm pretty fast myself."

Smiling one of the biggest smiles that I'd ever seen, he now glanced around the classroom, giving each of us personal acknowledgment. This teacher seemed so different from all of our other teachers. He didn't look tired. He just seemed to be bursting with vitality and happi-ness. "Jody and I were married," he said, sitting on the front part of his desk, "right after college. We both love sports. In Colorado, we loved skiing and here in Southern California, we love to get up at daybreak and hit the surf before coming to school. There's no feeling in all the world like coming down a slope in the Rockies, going forty, fifty miles an hour, except for feeling the wind and spray flying past your face when you catch a good wave. We both love speed! We plan to build a sailboat with our own hands, have two-point-five children, and sail around the world."

Some of us screeched! This was so funny—he and his wife were going to have two and a half kids! But I could also see that not all of us were laughing. Some kids didn't know what to make of him.

"There," he said, grinning to us from cheek to cheek, "in a nutshell, I've just shared with you what I love to do. Now I want you to share with me what you love to do," he said, getting to his feet. "And I don't care about spelling and punctuation. What my wife and I found out when we were ski instructors in Colorado was that the most important thing was to get our students on the slopes and get them all excited about skiing. Technique, we found out, comes later. Fun and excitement come first, then people learn a hundred times faster.

"Get it? It's not normal or healthy to keep energetic young kids like you closed up in a classroom. By nature, you want to play and run and jump and have fun, so that's why I don't care about periods or commas or misspellings at this time, even though this is English class. What I want to do is to get you so excited about reading and writing that the love of learning will be with you for the rest of your lives. I don't want to hinder your natural joy of wanting to venture into a world of books and writing and wanting to learn." He stopped and looked around at all of us. "Who likes what I've just said?" he now asked with a big grin.

I shot my hand up, and I thought that everyone else had shot up theirs. But I was wrong. A few cadets didn't raise their hands.

"All right," he said, "for you that don't feel comfortable with this method of teaching, I'll keep you off the slope, and grade your papers for grammar and punctuation, too," he added.

I found this very interesting. The cadets who hadn't put up their hands were our three regular A students and one of our C students. Myself, I was mostly a D student, but I also got some Cs and C-pluses now and then.

"All right," he said, "start writing. And I repeat that the key point is, I shared with you what it is that I love to do, so now I want you to share with me what it is that you love to do. And I don't care about what it is. It can be about anything. All I want is to be able to FEEL the EXCITE-

MENT of what it is that you love to do!" he added with such vigor, that I thought he would burst.

Hearing that I didn't have to worry about misspellings and punctuation marks, I was given wings. My God, I'd never heard a teacher call himself a bum before, and then share with us stories about his own life with such passion. Quickly, I started writing about what it was that I had always liked to do the most when I'd been a kid. It was going horseback riding in the hills behind our home, taking my trusty Red Rider BB gun or my big, tall bow and arrows, so I could hunt squirrels and rabbits. But I could see that some of my fellow cadets didn't start writing right away. One of them wanted to know how much time we had. Another wanted to know how long our papers should be.

Mr. Swift told us that we had the whole session to write, and our papers could be two to ten pages long. That he'd collect our work at the end of the period, grade them tonight, and give them back to us tomorrow.

Our top A student, Wallrick, who also happened to be our class leader, wanted to know if we'd get higher grades if our papers were closer to ten pages than two.

"Yes and no," said Mr. Swift. "I'll be grading on quality first, but quantity will also influence me. A hundred-yard dash—like my wife used to run—is a thing of beauty, but a full marathon is—" he added, shaking his head. "—beyond belief!"

They continued talking, but I was already writing, so I didn't pay much attention to them anymore. I then began to write about my big, older brother Joseph, who we'd always called *José* or *Chavavoy* in Spanish at home, and I wrote that he'd died when I was in the third grade. That was the first time that I flunked a grade.

Finally, I heard our substitute instructor tell Wallrick, who continued asking questions, that discussion wasn't always a sign of an inquisitive mind, but could also be interpreted as procrastination. Then he told us to all get started, right now.

I was on fire. I was already into my second page. My God, without the shackles of spelling and punctuation, I was flying! All these things were coming out of me. Now, I was writing about how my big brother Joseph had been sick and in the hospital for a long time before he'd died. My parents were gone a lot of the time, visiting him down at Scripps Hospital in La Jolla, so I was very happy that my brother's dog Shep had started to spend a lot more time with me. Because, you see, dogs, like cats and horses, wouldn't spend time with you unless they really liked you.

I'd been eight years old at this time, I wrote, and in the third grade, and every afternoon after school, I'd go hunting with my bow and arrows and my brother Joseph's dog Shep. Shep had been on the ranch when we'd moved from the *barrio* of Carlsbad to South Oceanside. Shep wasn't very big. He was half coyote and half sheepdog. And when we'd go hunting behind our ranch up the railroad tracks that ran inland from Oceanside to Vista, Shep never chased the rabbits or the squirrels like other dogs. No, he'd watch which way the squirrels or rabbits were headed, then he'd race around, circling them real fast, beating a few of them to their holes. And those that didn't make it to their holes, he'd chase up a tree or fence post.

Then—I saw him do this hundreds of times—Shep would get under the old tree or fence post, make eye contact with the squirrel, then he'd begin to circle with his tongue hanging out, salivating. That squirrel up in the tree or on a fence post would get so nervous watching him circling about below that he'd turn around and around, trying to keep his eyes on Shep.

Then suddenly, Shep would change his direction and run really fast around the tree or fence post in the opposite direction and that old squirrel would now get so dizzy that he'd fall out of the tree or off of the fence post. Instantly, Shep would be on that squirrel, expertly grabbing him by the back of the neck and giving him one or two shakes with such force that he'd break the squirrel's neck, killing him. Everytime we'd go hunting, Shep would get more squirrels than I could with my BB gun or bow and arrows. In fact, if the truth be told, my BB gun was so weak that the BBs wouldn't even penetrate the skin of a rabbit or squirrel unless I was super close, and with my big long bow I was still so slow and clumsy, that when I was in the third grade, I never got any game.

I then concluded my paper by saying that for months, I'd hunted with my brother's dog Shep before and after school, but then when my brother died, Shep ran off into the hills, never to be seen again. I couldn't believe it, I'd handwritten nine and a half pages. Boy, I'd been flying down that slope at a hundred miles an hour! But I didn't want to stop, even though class was over, and Mr. Swift was now collecting our papers. I still wanted to tell our teacher about how the Indian people who'd worked on the ranch for us had explained to me that Shep, who'd always loved my brother more than life itself, had disappeared, because he'd run off to the highest hilltop to intercept my brother's soul so he could lead my brother's soul back to heaven.

Dog, cats, horses, all animals could do this at will much easier than us humans, Rosa and her husband Emilio, had explained to me, because animals were much closer to God than we humans were. This was why Jesus had been born in a manger, to learn about love from the animals in the barn so that He could then do His Most Holy Earthly Work when He got big.

I turned in my story, entitling it "The Smartest Human I Ever Met, My Brother's Dog, Shep," and the next day, when I got my paper back, I had an A.

Instantly, I knew that there was something really wrong with this substitute teacher. I'd never gotten an A in all of my life, so I took Mr. Swift aside to speak to him, so he wouldn't get in trouble.

"Look," I said to him, "you might know a lot about going down cliffs of snow and surfing, sir, but I don't think that you know too much about teaching."

"Why do you say that?" he asked.

My heart was pounding like a mighty drum. I knew that I'd be cutting my own throat by saying what I was going to say, but I had to. I didn't want Mr. Swift to get in trouble. I liked him. I really did. "Because, you see, sir," I said, my *corazón* beat, beat, beating, "I'm a Mexican, and so you can't be giving me an A. You got to give me a D or C, or maybe you can get away with giving me a *C+.*" Tears burst forth from my eyes, "Because I've had a few of those, but you can't," I added, feeling so tight in my chest that I could hardly breathe, "give me an *A.* That's wrong, sir."

He looked at me like he didn't know what I was talking about, took a deep breath, and gently put his hand on my shoulder. "I don't know what you're talking about," he said. "All I know is that I never saw animals as being so smart and special until I read your story."

"Really?"

"Yes, really. And I'm now also convinced that your brother's dog Shep is the smartest human I've ever met. Tell me, is it part of your Indian upbringing to refer to animals as being human?" he asked.

I shrugged. I didn't know how to answer his question. It was still very hard for me to sometimes know where my Catholic–Christian upbringing stopped and my grandmother's Indian teachings began. For me it was all like one big river running together with all these different waters. By the time our local San Luis Rey River got to the sea, who could tell which water had come out of which canyon, especially when you realized that the San Luis Rey River didn't just start at our towering Palomar Mountain some thirty miles inland, but actually started beyond Palomar in all that huge, open eastern country, where my dad and my uncle Archie liked to go deer hunting with Archie's relatives from the Pala Indian Reservation.

"Okay, you don't need to know," he said, "just keep writing as pure and honest as you wrote yesterday. I'm looking forward to reading what you'll write today. The *A* remains."

I was double, triple, quadruple SHOCKED! I'd never had an *A* in all my life! My heart was beat, beat, beating a million miles an hour! I wrote that day in class as I'd never written before, expanding on the story that I'd written about Shep and how on the night of my brother's death, Shep had howled all night long at the moon, and then he'd disappeared in the early-morning hours of the night.

Then I wrote that when our dad and mom came home that morning from Scripps Hospital in La Jolla and informed my little sister Linda and me that our brother Joseph had died, we weren't surprised, because Rosa and her husband Emilio had already told us. I began crying as I wrote. Mr. Swift came by and stroked me on the shoulder when he saw me crying. Class was over, but I'd been so far away in my writing that I hadn't even realized that class was over, and the rest of the cadets were getting up so we could be marched to our next class. My God, Mr. Swift had been right; once we got out of the confines of spelling and punctuation, writing and reading could be so exciting!

Sleeping, dreaming here in my hotel room in Long Beach, I remembered all this like it had happened in a far away, foggy dream. My God, I really hadn't realized it, but I owed so much of my joy of reading and writing to that substitute teacher in the seventh grade who, in three tiny days, had touched my heart and soul.

My eyes filled with tears. My God, how I'd like to find that teacher and give him a copy of my book *Macho!* I breathed, and breathed again. Teaching didn't have to be long and boring and laborious. No, teaching could be done as fast as a lightning bolt. He'd cut across the valleys of my deepest doubts, giving light to the darkest crevices of my beaten-down, inhibited mind, accessing a natural storytelling ability within me that was utterly profound! Oh, how I wished that I'd told those teachers yesterday about this Mr. Swift, or Mr. Smith—I couldn't really remember his name—but I knew that it had started with an *S*, followed by an All-American-sounding name.

Wiping the tears out of my eyes, I stretched and breathed and I woke up some more, hoping to God that I could at least find that man from the teachers' union today and tell him that yes, he'd been right, and I'd found a wonderful teacher in my life who'd helped me along my way.

I stretched some more, rubbing the sleep out of my eyes. I could now hear the quiet surf of the Long Beach Harbor outside of my window. I was on the tenth floor, and I could see that the light of the new day was just beginning to paint the eastern sky with colors of orange and yellow and streaks of pink. It was going to be another gorgeous day . . . here in the paradise of Southern California.

I could now very clearly see that that man from the teachers' union had been brilliant when he'd told me that "wake-up talks" were good, but that teachers also needed to know what worked in our system. Yesterday, when I gave my talk, I'd been so hot with rage that I hadn't been able to remember anything good that had ever happened to me in school. But now, this morning, after a goodnight's rest, I remembered some very good things, too.

I could now remember so clearly that this substitute teacher—who I'll continue calling Mr. Swift—gave me not just wings, but asked me some questions that would provoke my mind for years. I'll never forget that day when I walked out of the classroom with the first A of my entire life. Instantly, the real A students, who'd been waiting for me, knocked my books out of my hand and they grabbed my A paper away from me and started saying, "You didn't spell this right! You didn't use punctuation! How could he give you an A?"

"What did you guys get?" I asked.

They mumbled, but not very clearly, and crumbled up my A paper and threw it on the ground. I almost started to cry, but didn't, and when I bent down to pick up my books and my crumbled-up paper, I found two hands helping me. I looked up and saw that the two hands belonged to George Hillam. He was a day student like me, meaning that we got to go home every afternoon and sleep at our own homes. We didn't live on the school grounds like the other students.

"They got Bs," said Hillam to me. "That's why they're pissed. And Wallrick, he even got a D, and the teacher wrote on his paper that he knows where he got a lot of his story."

"You mean that Wallrick copied?" I asked quietly as we picked up all my books and other stuff.

"No, he didn't copy, but he used material that wasn't his."

"I don't get it," I said.

"Look," said Hillam very patiently, "we were told to write about what we loved to do, and he wrote about what someone else had written that they love to do, which was boating. Get it? He was trying to butter up our new teacher, like he always does with Moses and our other teachers."

"Oh, I get it!" I said, finally understanding. "Because Mr. Swift had said that he and his wife are going to build a sailboat, he wrote about boating, but used somebody else's story, thinking that this was the best way for him to get a top grade out of our new teacher."

"Exactly," said Hillam.

"What did you get?" I then asked Hillam.

"I got an A, too. I wrote about loving to bake cookies and eating them all!" said Hillam, laughing so hard that his big belly shook up and down.

George was fat and one of the most well-liked guys in our whole grade. His father owned a grocery store in downtown Oceanside and was real tall and skinny. It was George's mother, who had the prettiest blue eyes in all the world, who was heavy-set like George.

Everybody was already standing in ranks. George and I were the last ones to get in file. It seemed like George and I were almost always late for everything. Usually George only got Cs and Bs. He was pretty excited about his A, too.

For three days I got As and George did, too, and finally I even forgot that I was Mexican and not a good student, and all those other derogatory things that had been pounded into my head ever since I'd begun school. For the first time in my life, I was no longer one of the slowest learners in my grade. No, I was simply becoming a regular kid who loved to come to school, work extra hard on my homework at home, and could hardly wait to get to school in the morning so I could go to English class.

I was flying!

I was an eagle shooting across the heavens!

I was that Redtail hawk that lived in the towering Torrey Pines by our home and screeched with gusto every afternoon when I went hunting, causing flocks of dove and quail to take flight. And I now understood for the first time in my life why reading and writing had always seemed so boring and stupid. I'd never known that they were vital to our lives outside of the classroom. You see, Mr. Swift explained to us that the whole world communicated through reading and writing. A scientist learned about science through reading, then later in his career, he expressed his findings to other scientists through writing. A farmer learned about new tractors and the new methods of farming through reading, then he also told the world about any discoveries that he might have uncovered, through writing.

Years ago, Mr. Swift explained to us, it had all been done around the campfire—the sharing of stories and information—but now in modern times, with the known world having gotten so much larger, it was all done through the written word. The plumber, the electrician, the lawyer, the doctor, the trainer of dogs and horses—everybody learned and communicated through the written word.

And with a library card, he told us that we could learn the fundamentals of anything in all the world; then, with hands-on experience, we had no earthly limitations. This was all so mind-boggling for me. I'd always thought libraries were for punishment, to get us to keep still and shut up. I'd never thought of all those books in the library as our "gateway," as Mr. Swift said, to the whole universe of knowledge.

Now I was writing about Midnight Duke, the most noble horse we'd ever had on the ranch, who'd break down gates or climb over fences to be at the side of a mare who was going to foal, so he could guard her against the coyotes who lived below our house in the swampy area, where the sea came into our inlet of fresh water. But then, on the fourth day, when I came in to show Mr. Swift what I'd written at home for him, he was gone, and there was Captain Moses once again.

Our three A students quickly went up to Moses and started talking to him excitedly, causing a big commotion, then they pointed at me. I glanced at Hillam. Why weren't they pointing at him? He had also gotten As while Moses was gone. I could clearly see that Captain Moses was getting so mad as they spoke to him, that his ears were turning red, then they began to pulsate with a purple glow.

Suddenly, we'd no more than taken our seats, and Moses came rushing across the room straight at me and ordered me to get out of my seat and stand at attention. He began to yell at me and poke me in the chest with his index finger, so hard that it really hurt.

"Give me back your papers!" he ordered. "I don't know what you did to that teacher while I was gone, but you will not get away with this any-more!"

I quickly handed him my papers as I'd been ordered, including the one I'd just written last night about Midnight Duke and a mare giving birth. Moses took my papers and he went to his desk—not giving me permission to sit back down—and brought out his big red marker and began going over my papers with a vengeance.

I stood there feeling all alone and so terrible that I could cry. Another thing I'd noticed that Mr. Swift had done was that he'd graded all of our papers in pencil, and lightly, too, like he respected the work that we did. But Moses, he was now marking up my paper with an anger and frus-tration that I'd only seen with castrated geldings when they'd mount a mare, but they couldn't get anything going.

He was sweating by the time he finished marking up my first paper, and he now crossed out the A which had been penciled in, laughed, and put a huge D across the paper, and said aloud to all the class, "He's such an ignorant farm kid that he thinks that animals are human and can think and have reason."

All the class laughed at me, especially the three A students. The only one who didn't laugh at me was Hillam. He was looking down at the floor and I could see he was feeling real bad for me.

Having given me a D, Moses now went after my second paper, and this time he crossed out the A before he'd even read my paper. Then, after going over this paper, he laughed and told the class how I'd written about my dog Shep disappearing to lead my brother's soul to heaven when he died. He then told everyone that science had proved that there was no such thing as a soul. And also that he'd personally known my brother Joseph when he'd attended the Army Navy Academy and that Joseph Villaseñor had been a real cadet and not afraid to go out for football like I was, and that I was a little sissy Catholic who liked to wear a dress and help the priest say mass.

I lowered my head. This was awful! My mother had gotten me a special pass to leave school early on Fridays so I could learn to be an altar boy, but I'd failed. I hadn't been a good enough reader to be an altar boy, either.

Moses now made this second D even larger than the first, with his red marker. In fact, I could see the D from clear across the room where I was standing at attention. I had tears running down my face, but I was afraid to wipe them for fear that I'd get into even more trouble.

Then Captain Moses took hold of my third paper that I'd handed in, the one on which I'd written my full name, Victor Edmundo Villaseñor—when normally I only wrote Victor Villaseñor—and the first thing he did was lash a line across Edmundo, wrote E-d-m-o-n-d in huge letters, and laughed again, this time telling the whole class that I was so stupid, I didn't even know how to spell my own name, and he gave me a huge F without reading any of this paper.

I remember very well that something snapped inside of me at this point. I don't know what it was, but I pulled back my shoulders and stood up as tall as I could, happy that I wasn't seated, and now I stared at Moses, no longer ashamed of the tears that were running down my face. Because these tears, they no longer flowed out of my eyes with fear, but with pure-white, GUT-GRIPPING RAGE!

I was a cadet no longer.

I was now a warrior, an ancestral brave, hearing the beat, beat, beating of my heart as I'd never heard my heart beating before. Never again would I ever pee in my pants at school, or at night when I slept, because . . . simply, I now knew that I was going to kill . . . to kill this teacher Moses who stood before me, just as I'd helped kill well over a hundred head of livestock on the ranch in the last ten years.

I was no longer in the third grade and hunting in the hills behind our home with a BB gun or a bow and arrow set that was too big for me. I now hunted birds with a .20 gauge shotgun, rabbits and squirrels with a .22 rifle, and I hunted deer with my dad's 30/30 Winchester lever-action saddle gun. In fact, last year at the age of thirteen, I'd taken and passed my hunter's safety course, gotten my first California state hunting license, and I'd taken my first buck with my dad's rifle with a three-hundred-yard running shot.

I was excellent with weapons and horses! Absolutely excellent! And I also knew how to track and read terrain better than anyone I knew. No one in all my school had ever hunted or taken as much game as I'd taken. And so, looking at this man in uniform before me, I now knew that never again would I ever get so scared of him . . . or anyone else that I'd feel like pissing.

No, now I was set. I was in that smooth, all-true place that my dad had explained to me that every good man had to get to so he could defend his home, his *casa*, his *familia*, as my dad had done with his mother and sisters in the Mexican Revolution.

My heart was beat, beat, beating like a mighty drum, and the tears were running down my face, but I wasn't the least bit scared or nervous anymore. No, I was now simply at this place called a Mexican stand-off, or *un gallo de estaca*.

Moses was now shouting at me, ordering me to sit down, but I no longer heard him. No, I was good where I was, right here, standing on my own two feet. I was excellent, in fact, here at my own safe spot, and I'd never be messed with again. Because, simply, from now on, if anyone pushed me or insulted me, I'd kill them . . . just as I'd killed that deer.

And why? Because, simply, a person's name was sacred. Given to us by people who loved us. And my name E-d-m-u-n-d-o had been given to me by *mi papa*, who told me that he'd given me this name because one time when he'd been in prison—my dad had gone to prison three times—a big, huge cook, called *Patas Chicas*, meaning Little Feet, because his feet were so monstrous, had taken my dad, a Mexican-Indian like himself, under his wing and tried to teach my dad how to read, explaining to my father that reading and writing weren't just White People's tools, but also the future for *nuestra gente*.

The huge cook had read the book called *The Count of Monte Cristo* in Spanish to my dad. That book's hero was named Edmundo Dantes and he'd been sent to prison unjustly, but he escaped, found a treasure, and was then free to avenge himself on all those who'd put him in prison and stolen his bride-to-be. But, then, Edmundo had found a light even higher than revenge, and that had been *amor*. A Frenchman named Dumas had written that book, and that Frenchman—Little Feet told *mi papa*—had been a black man, the son of a slave straight from Africa, and on his fiftieth birthday, that gigantic black man had a harness made for him out of leather that he'd put over his shoulders, and he'd then gotten up in the hayloft over his stables and raised a draft horse one foot off the ground with his legs and back muscles. This, my dad told me, was the story that had given him, *mi papa*—the guts, the wings, the hope with which to survive prison—and he, just being a boy of thirteen, and all those monsters in prison trying to rape him.

"And so I named you Edmundo," my dad told me with tears in his eyes, "to give you the strength *de corazón*, to never give up, no matter what twists life throws at you or how impossible it all might seem at times. You are Edmundo! Victor Edmundo! The victor over all odds just like Edmundo Dantes!"

And here, this teacher, Captain Moses, had ridiculed my name, a sacred name, changing it into something else, with no respect for me or *mi familia* whatsoever, just like what had been done to the slaves that had been brought over from Africa and the Indians who'd been run out of their sacred lands and put on reservations.

Yes, with all my heart, I was going to—not just kill Moses—but gut-shoot him, then rope him and pull him upside down into a tree like we did to the livestock that we butchered. Then I'd cut him open from rib to rib, making sure he didn't die, so he could watch his intestine slip out of his middle, and then he'd scream when I poured a sack full of live rats over him to eat him alive.

Now, I was the one who was smiling a nasty little smile as I stood there in class, having these juicy good thoughts of revenge. Captain

Moses continued shouting at me to sit down. "THAT'S AN ORDER! Take your seat, cadet!"

But I didn't. I just stood there, calmly looking at him. I was free. Never again would I let myself get trapped and enslaved with fear like had happened to me since the first day that we'd started school and they'd slapped Ramón, the best and bravest and smartest of us all, until he'd been bloody, terrifying the rest of us.

I breathed as I looked at Moses yelling at me and I felt as strong and free as when I went home and passed through the gates of our huge *rancho grande,* and I knew then that nothing bad could ever happen to me again because I had *familia* . . . and their blood pounded strong in my *corazón!*

I was set now!

I was free!

I'd found my place, this warm secure place from where I could look out on the world without fear. Never again would I get so scared that I couldn't hold my pee.

The second alarm went off. Class was over. I'd never sat down. I'd beat Moses. And everyone had seen it, too.

I picked up my books. I was the last one to go out of our classroom. All of my classmates were already outside assembling, so we could be marched to our next classroom. Suddenly, as I walked outside, our three A students rushed at me, hitting me with their shoulders so hard that I was knocked to the ground. My books and notebook went flying. They gave me a few quick kicks to the ribs, then slapped fives with one another. I couldn't breathe. They'd really kicked me hard. Quietly, I picked up my books and notebook, but strangely enough, I could see that, well, I wasn't really feeling as intimidated or scared as I normally would have been. George Hillam came over, but he didn't help me pick up my things this time.

"You fool," he said under his breath to me. "You've really done it this time. Don't you know, you can't fight the establishment. You can have those thoughts. We all do. Even with our parents. But you can't be staring at people straight in the face, telling them with your eyes that you want to kill them, without them getting you."

I was all confused. Hillam was always so quick to understand everything, like so many other fat people that I knew, that he always confused me.

"Moses told them to get you," he added. "That's why I can't help you. They'll get me, then, too."

"Oh," I said, looking behind me and seeing that Moses had watched the whole thing, and that he was now putting on his military cap and walking off to be with the other teachers at the head of our assembly. "Now I get it," I said to Hillam, "then this is war."

So now realizing the situation, I quickly finished picking up my things and tried to think about all the military maneuvers that we'd studied in our military history courses. They'd really hurt me bad. And it was Wallrick who'd kicked me the worst.

I took a big breath, glanced around, and tried to figure out a way to avenge myself. I spotted the new blond cadet named Igo, who was from North Hollywood, and his mother or uncle or someone in his family was in the movie business. Igo was still so new that he didn't have any friends yet. He was a big, husky, strong-looking kid. So I got next to him when we lined up to be marched to our next class. We were all standing at attention to be counted off, when I poked Igo in the ribs, and instantly got back to attention with my eyes straight forward.

"Stop it!" said Igo, and he turned and poked me back.

Hearing the commotion, Wallrick, our top A student and class leader, turned and saw Igo poking at me.

"Knock it off, Igo!" he barked.

"But he poked me first!" said Igo.

"Did you poke Igo first?" asked Wallrick of me.

"Not me, sir!" I said, lying, and pretending like I was so scared that I'd never lie or misbehave. "He's a lot bigger and stronger than me, I'd never do that!"

Feeling satisfied that I was a coward and knew my place, Wallrick turned back around to present our class to the extended assembly of cadets and instructors.

Quickly, I poked Igo again, sticking my tongue out at him this time. Igo, a big easygoing guy, laughed and hit me on the shoulder. And Wallrick, true to form, turned and saw our commotion again, and realized that the other assembly leaders were waiting for him to bring his class to order. He got so mad, he came racing into our ranks and got in Igo's face.

"YOU SCREW UP ONE MORE TIME, IGO!" he barked. "And I'll see you before messhall behind the barracks to teach you how to follow orders!"

"But sir, he's the one who—"

"Don't 'sir' me! I'm a sergeant! I work for a living!"

"But Sergeant, he's the one who's doing all the—"

The eyes of our A student leader were bloodshot with rage. He glanced at me, but I was standing at attention with my shoulders back, looking completely military-proper and scared, so our brilliant leader was just sure that I was too chickenshit to be doing anything wrong.

"Don't be lying to me!" He turned, barking at Igo. "Meet me behind the barracks if you have any balls!"

"Well, okay, I'll be there, but . . . honestly, he's the one who—"

"ATTENTION!" barked Wallrick. "I will not have anyone in my squad making me look bad!"

I'll never forget—we went to our next class, but we couldn't pay attention because we were all so excited. Then we were dismissed to clean up before messhall, meaning lunch, and we all rushed behind our barracks to see if this whole thing was really going to go down. And there was Wallrick, stripped to the waist, and it was easy to see that he was the biggest, strongest kid that we had in the whole seventh grade. He had already started shaving, was almost six feet tall, and looked like he should be a junior or senior in high school, and yet he was only about thirteen years old like the rest of our class. In fact, because I'd flunked the third grade, I was actually about a year older than most of the kids in my grade, but I had such a baby-smooth face that nobody believed that I was as old as I was.

Wallrick was flexing his arms and cracking his knuckles. And Igo, he finally showed up, but you could see that he really didn't want to fight, and yet he'd do it to save face. The other cadets made a ring around Wallrick and Igo. Myself, I kept back out of sight because I didn't want Igo to see me and suddenly decide to come running after me, instead of going after Wallrick. Everyone was betting on Wallrick to win, but myself, I figured that Igo had a good chance because I'd horsed around with him a few days back and I'd seen how quick he was.

With fire in his eyes, Wallrick came swinging and cursing at Igo, telling him how he was going to teach him a lesson, because he wasn't going to allow anyone in his squad ruining his chances of eventually becoming the adjutant cadet of the whole school and then going on to West Point.

Igo hit him.

Igo hit Wallrick with a straight right to the face while our A student leader was talking about his future. It shocked Wallrick. In fact, it shocked all of us. Wallrick stopped and reached up to his lip with his right hand, then he looked at his fingertips, saw the blood, and his eyes went crazier than before.

Bellowing, he now went rushing in on Igo again, but the kid from North Hollywood was really fast, so he just sidestepped and hit Wallrick with two more quick lefts to the face, then a right uppercut to his middle. We all groaned with the sound of the hit to the stomach. Igo really knew how to box, but Wallrick was bigger and stronger. He wasn't our class leader for nothing, so he didn't go down. No, he caught his breath and now lunged at Igo, got him in a headlock, and started twisting Igo's head off of his neck with all his might, and then began kneeing him to the body.

Igo cried out in pain, but he was tough, too, so he now lifted Wallrick off of the ground, threw him down, and now they were both pounding on each other as they rolled around on the grass. My God, I'd never meant for all this to happen. I'd just been trying to even the scorecard a little with Moses and our A students, who'd knocked me down and kicked me.

Well, the fight continued until they were both a bloody mess. But you could tell that Igo had won, and now he had lots of friends. At this point we all had to go to our rooms to clean up before we could go to lunch. But when Igo saw me, he forgot all about lunch and took off after me, saying he was going to kill me. I had a head start, so I took off running and hid behind the trash cans, where it smelled so bad that I figured that he wouldn't want to search for me very much.

It worked. He left. But then, right after lunch, Igo spotted me again and this time I didn't have much of a lead, so I knew that he was going to catch me because I wasn't that fast of a runner.

I ran down alongside a building, trying to figure out what to do real quick. Then I remembered how my dad had taught me to set traps for rabbits as he'd done in the middle of the Mexican Revolution to get food for his mother and sisters who'd been starving.

Rounding the far corner of the big building, I pulled out my pen and pencils from my book bag, selected one pencil because it was recently sharpened and held it out like a spear as I heard Igo coming around the corner at a full out run. And he was fast, too, so he hit the end of my sharpened pencil with his chest at full speed, which drove the pencil into him.

Seeing the big, long yellow pencil sticking out of his chest—I'd let go of it the moment I felt it penetrate—he screamed in terror and started turning pale and began to gasp.

"You killed me!" he screamed. "You got me in the heart!"

"No, I didn't," I said, having butchered a lot of livestock. "Your heart is over here, more on your other side. Besides, I never drove it. I just held it, and you ran yourself into it just like a rabbit running on a trail into a trap of sharp sticks."

"I'm going to get lead poisoning," he said, looking weak with fear.

"I'll yank it out," I offered.

"No!" he screamed.

But I'd already grabbed it and yanked it out. He fell down crying. "Get me to the infirmary," he begged. "I don't want to die."

"You won't," I said. "I've poked myself with a pencil before and I didn't die."

"But I can get lockjaw?" he said. "Come on, help me up. I need to get to the infirmary."

"All right, I'll help you up, but no more trying to get even with me. I'm sorry that I got you into that fight with Wallrick, but I never expected it to go that far. I'd just wanted to get a little bit even with him for knocking me down and then kicking me. Besides," I added, "you beat him, so you'll never have any more problems here at school. Hell, you should be happy with me. You're a hero now."

"You sick bastard," he said. "You had the whole thing planned since that first time you poked me, didn't you?"

I smiled, but said nothing more.

Then, getting to the nurse's office, Igo did something that I'll never forget . . . he didn't turn me in for having stabbed him or setting up the whole fight. He simply told the nurse that he'd fallen on his own pencil. We became good friends. And Moses, when he found out that Igo had beaten the crap out of his top pet A student, and that Igo and I had become friends, he didn't like the smell of the whole thing, but he didn't send his bully pets after me anymore.

Still, I didn't win the war. Only a few battles here and there. George Hillam was right, you couldn't fight the establishment. And Moses just kept at me, giving me as bad of grades as he could and ridiculing me in front of everyone, until finally he had me so beaten down once again that I actually began to hate Mr. Swift for ever having given me any hope.

I was STUPID! There was no getting around it!

And spelling and punctuation were what mattered, not what you dreamed or loved to do!

Still, as sure as the sun went down and the moon came up at night, it was now written in my heart and soul . . . Moses was a dead man.

CHAPTER three

Getting up that morning in Long Beach, I never found that man from the teachers' union. They said that he'd left, but I did get so many requests for speaking engagements that Karen Black, our publicist, told me that my New York publisher was starting a lecture agency for their authors and that they could do my bookings for my talks for 25 percent of my fees. I said that I hadn't even realized I was going to be paid, so I'd have to think about it.

Then I found out that the most prolific living writer in all the world, the Western writer Louis L'Amour, who had more than 300 million books in print, was going to give the talk at today's luncheon. His book *Hondo* was one of my all-time favorite books, just like it was one of my all-time favorite Western movies, and Steinbeck, Faulkner, James Joyce, Anne Frank, Tolstoy, Dostoevsky, Azuela, Rimbaud, and Camus were some of my all-time favorite writers.

I met with Louis L'Amour and his wife Katherine and their two kids just before his talk, and I asked him if he had any advice for a new writer.

We found a quiet place in the bar but didn't order drinks, and he said, "First, a writer writes. That's his job, his work. Second, it's never about money. It's about writing, that's why I didn't get married until I was fifty, because I didn't make enough money to be married and also keep writing. And about my book *Hondo* that you say that you love, I sold it to the movies for five hundred dollars and people tell me that I was cheated. I say, no, I needed those five hundred dollars, and they paid me. Movies are free advertisements for our books. I heard that you caused quite a stir yesterday with your talk. That's good, you got their attention. Now go home and write.

"The thing that ruins more writers than anything is drinking, thinking about the money, wanting to hobnob with the rich and famous, and going to too many conventions like this one. John Wayne has been in three movies that were made from my books, and I've had the chance to meet the man several times, but I never have. Why? Because, simply, writing is a job like digging a ditch, or building a house, barn, or corrals, and I was too busy writing at those times to meet the man.

"Early success has ruined more of our best writers than anything else," he added. "They write an excellent first book because they had plenty of time and nothing to lose. But then, with the taste of success, they freeze up or start drinking and hobnobbing, or thinking that they should go into politics, and they never get a second book out of the same quality or intensity. So, keep your powder dry and dig in for a long, fruitful life of being a writer, that storyteller around the campfire of your people and your generation. Your trade is as old as time, and your main job is to uplift the human heart so that then we can go on with dignity and fair play. That's it."

I was stunned. I'd never expected to be given so much, and all in a nutshell. He, then, told his wife to give me their home number and he told me to call them anytime, day or night, because he'd read the first few pages of my book *Macho!* and he could tell that I was the real thing, as the author Hemingway always liked to say.

I got in my Ford van, my "white stallion," and drove home, feeling ten feet tall and ten years older than when I'd driven up to Long Beach only thirty-some hours ago.

A few days later, the Long Beach paper came out with a big picture of me and a long article, then the LA *Times* ran a feature story on me, too. A few weeks later, I couldn't believe it, the *Times* also reviewed my book, comparing *Macho!* to the best of John Steinbeck, the great writer himself!

Oh, I was on my way! I had to breathe real easy just to keep calm— I was flying SO HIGH!

BOOK **t w o**

CHAPTER **four**

I was five years old. The year was 1945. My father, Juan Salvador Villa-señor, walked me from our old ranch house up to the corrals to talk with me.

"*Mijito,*" he said, "tomorrow you'll be starting school, so it's important for you to understand who you are and who your people are. You are *un Mexicano.* And *Mexicanos* are such good, strong people that everywhere, everyone wants to be a Mexican. Look at what used to happen back in the *barrio* in Carlsbad, the *gringos* and the *negros* would come to our poolhall, eat a few *enchiladas,* drink a couple of *tequilas,* and then they'd start to sing with the *mariachis.* That's proof that everyone loves the Mexicans, and wants to be *un Mexicano.* Get it?"

I nodded. "Yes, I get it, *papa.*"

"Good, because now that you're starting school, you have to be a good little man and start studying the girls, so when your time comes, you'll know how to choose the right wife. Because the most important thing any man can do in all his life is pick the right woman to breed with—I mean, marry first, then breed, because from the woman comes the—"

"—comes the instinct to survive," I said, having heard this for as long as I could remember.

"Good," said *mi papa,* "very good. You remembered. And so for you to be attractive to girls, *mijito,* you can't be picking your nose anymore and wiping it off on your pants. Do you understand? *Lo cortés no quita lo valiante, y lo valiante no quita lo cortés.*"

This I'd also heard for as long as I could remember, and it was one of our oldest Mexican *dichos,* sayings, and what it said was that manners didn't take away bravery, and that bravery didn't diminish manners.

"Yes," I said, nodding, "I think I do."

"And also," said my dad as he continued smoking his big, long cigar and we passed under the huge old pepper tree, "from now on you got to be responsible, and this starts with every man and woman knowing how to wipe their own ass."

I nodded again. I was listening real closely to every word that *mi papa* was telling me, because growing up on a ranch with horses and cattle and big trucks and tractors, I'd learned that if you didn't pay attention real close to what you were told, the next thing you'd know, you'd be run over by a tractor, or be on horseback and have your saddle slip out from under you, or worse still, you'd have a rattlesnake scare the living shit out of you because you hadn't been paying attention to where the Father Sun was in the sky and been watching out for the shady spots on the trail. But still, I was having a hard time listening to my dad, because my brain just kept jumping around inside of my head.

Hell, I'd never been away from *mi familia* before, and why did I need to go to school anyway. I was learning everything that I needed to learn there on the ranch. I knew how to milk a cow to get milk. I knew how to plant and grow corn so we could make *tortillas.* What else was there for me to learn?

"So are you understanding me, *mijito?*" my father now said to me, puffing on his cigar. "You're going to have to now know how to be out on your own."

I shook my head. "No, *papa,* I really don't understand," I said in Spanish. I didn't know any English. All we ever spoke on the ranch was Spanish. "How can I stop picking my nose? When my *mocos* get dry"—*mocos* means snot in Spanish—"and begin to itch, they hurt if I don't pick them. And my ass, I've never really figured out how to clean it real good yet. Do I bunch the toilet paper together, *papa,* or do I lay it out on the floor and fold it real carefully so that it stays flat when I wipe myself."

"Who showed you to lay it out flat and fold it?" asked my father. "I never thought of that. I've always just bunched it together. My God, *mijito*, look at you, you haven't even started school yet and already you've come up with a very good idea. I tell you, you're going to do good in school! Hell, you're already thinking, and that's what education is really all about, learning how to think."

Well, I felt good hearing this, but still, I didn't like the idea that I was going to have to go to school. "Can I at least go to school on horseback?" I now asked. I'd been riding horses since I was three, and on top of a big horse I felt like Superman, faster than a speeding bullet and stronger than a locomotive!

"No, I don't think so," said my dad.

"Why not? Uncle Archie said that when he went to school, half the kids on the reservation went on horseback and they got to take their rifles, too, so they could hunt for game on their way home for supper."

My dad pushed back his Stetson and scratched his head. "That was a long time ago, *mijito*. We can't just go riding or carrying guns into town anymore. We're civilized nowadays."

I really didn't like hearing this. I figured that I'd have a hell of a lot better chance at school if I could take my horse and my trusty BB gun rifle. On foot, I was still pretty damn short, and going into any new territory, I'd found out that I had a better chance if I went on horseback and was well armed.

I could hardly sleep that night—I was so nervous. I kept tossing and turning, and my older brother and sister were no help, because I'd learned so far in life that you had to take your own lumps when a horse threw you. Nothing anybody could tell you about getting bucked off could prepare you for the first time you ate dirt and felt so stunned that your brain couldn't even work until you'd taken a few breaths. But then, after eating dirt two or three times, getting bucked off wasn't all that bad anymore. I'd found this out first hand.

Monday morning, I got up extra early, washed, brushed my teeth, pulled up my bedding, laid out my clothes, and put on my new Levi's and the new long-sleeve checkered shirt my mother had gotten for me at JCPenney in downtown Oceanside. I loved going to Penney's with my mother, because at Penney's, they had a jar attached to a wire that they'd put your money in when you paid for something and the jar was then pulled on a wire real fast up to a little window above the store floor. The jar would then somehow miraculously slow down just as the window-woman reached out, took the jar, opened it, took out your money, made your change, wrote out a receipt, and then put everything back in the jar and pulled the cord and the jar came flying back down on the wire as fast as a bird with its ass on fire. Also, I loved Penney's because—like my mother Lupe always said—our pennies went further at Penney's than they did at Sears. But still, Sears was where we got most of our farm and horse equipment.

After breakfast, my mother took me into the bathroom and wiped off the egg that I'd gotten on my new shirt. I could now see that my mother had been very smart in insisting I get a checkered shirt instead of the plain blue one that I'd wanted, because the wiped-wet-area where she'd cleaned on my skirt was hardly even visible with all of the checkers.

When my mother was done cleaning me, she left, and I stayed behind in the bathroom alone. I peed in our toilet that was stained all orange from our hard well-water, then I got up on my little box so I could see myself in the mirror over the sink—which was also stained orange—and I saw that my hair was just about all combed down except where it always stood straight up in the back like the quills of a porcupine. Standing on the box, I made the sign of the cross over myself and began talking with God.

"*Papito*," I said, "You might have forgotten, because You're so busy and all, but today I'm going to school all alone and I'm just a little kid, especially when I'm on foot, so I'll need for You to please stay by my side and help me out in case I do something dumb and get in trouble. Okay? Do we got a deal, *Papito*, You'll stay by my side, eh?"

Making my request, I now closed my eyes real tight like *mi mama-grande* Doña Guadalupe had taught me to do, so I could then hear the voice of God inside of me. But what I heard next, I don't think was the voice of *Papito*, because now I heard my mother shouting, "Hurry up! You don't want to be late on your first day of school!"

My heart started pounding. Quickly, I made the sign of the cross once again, and said, "See You at school, God," and ran out of our smelly old bathroom, past the kitchen, out the back door, and to our car. I opened the car door, pushed the chicken off the passenger's seat where she'd no doubt decided to nest, and my mother and I were off, with chicken feathers flying all over us. Usually my mother went to work at about this same time, but this morning she was going to drop me off at school, then go to downtown Oceanside to do her bookkeeping at our main liquor store, which was located just up from the train station, near the pier. I loved the Oceanside pier. This was where Uncle Archie would sometimes take me fishing.

My mother drove us out from under the two huge pepper trees, passed our grouping of torrey pines, and drove down the long driveway of our huge *rancho grande,* underneath the umbrella of tall eucalyptus trees.

"You're going to love school," said my mother. "Going to school with your godmother Manuelita back in *La Lluvia de Oro* were some of the best days of my life."

"But weren't you a little scared on your first day, *mama?*"

"Yes, I guess I was, but your godmother Manuelita was the teacher's helper and she walked with me to school. Don't worry," she added. "You'll make friends, and then with friends, school and life are much easier, *mijito.*"

I hoped that my mother was right, because living on a ranch, I didn't know any kids my own age, much less have any friends. I guess that my dog Sam had been the closest thing that I'd ever had to a friend, but, then, he'd gotten run over by our German friends, Hans and Helen Huelster, about a year back.

I was looking out our car window as we drove. I could see that there were dozens of wild ringneck pheasants in our orchards of lemons and oranges. Seeing these beautiful birds gave my heart wings. Then we passed the dark orchard of huge avocado trees with the one old loquat tree where hundreds of birds liked to hang out. I laughed at seeing all the birds. Up ahead, we came to California Street. This was where we had our mailbox. Here we turned right, went past my aunt *Tota*'s house—she was my mother's older sister and married to Uncle Archie—then we took a curve in the road to the left, then a sharp right-hand turn, went past the new Hightower's Market, and came to Coast Highway, which was back then called Hill Street. At Hill Street we turned right. Hill, back then, was the biggest, widest, longest street in all of Oceanside. In fact, Hill Street was then part of the old 101 Highway, which ran up and down the whole coast of California.

Now my mother, a very good driver, speeded up and we went down a small hill, over the little bridge with the inlet of seawater that went up towards our house, up a short hill, across the railroad tracks, alongside the cemetery where my *mamagrande* Doña Guadalupe was buried, and then passed Short Street, which would later be renamed Oceanside Boulevard. Here was where I lost track of all the streets we passed. Because from here on, we weren't going along the outside perimeter of our big *rancho grande,* and so I didn't know my way around anymore.

Suddenly, up ahead, we turned right again and climbed a steep hill with lots of short blocks, as if we were going up to the great big Oceanside High School. And I was really glad that we'd made all right-hand turns, because I just didn't like left-hand turns. I remembered very well that my grandmother, when we'd lived in the *barrio* in Carlsbad, had only made right-hand turns when she'd push me in my stroller around our block when I was little, and so to this day, I still only felt good with right-hand turns.

Then I couldn't believe it; while I was thinking of my grandmother, my mother made a real sharp left-hand turn, and parked. My whole world felt like it had gotten all twisted around inside of my brain. Quickly, I flashed on *mi mamagrande* and my old dog Sam, and I just knew that I was going to need both of them, plus *Papito Dios,* if I were to survive the day.

"This is your school," said my mother to me, opening her door and getting out of our car.

I didn't like the look of things. There were kids running around all over the place and I didn't know any of them. My mother closed her door and came around the back of our car and opened the car door for me.

"Come on, *mijito,*" she said.

"No," I said. "I don't want to go, *mama.*"

"But you got to," she said.

"Why?" I said. "*Papa* didn't go to school and he always says that all a person has to do in this country is die and pay taxes."

She laughed. "Well, that's true when you grow up, *mijito,* and go into business, but right now you're still small, so you got to go to school before you can pay your taxes and die. Come on," she added, "give me your hand. I'll walk you in."

I still wouldn't give her my hand. "*Mama,*" I said, "can you please stay with me for my first day of school?"

"I don't know if I can do that," she said. "We'll ask. Maybe I can. Then I'll just go to the store, check the cash register, and come right back."

"Really?"

"Yes."

Hearing this, I felt a lot better, so I took my mother's hand and got out of our car. There were kids and parents rushing by all around us. My mother squatted down on her heels of her pretty red shoes and started picking all the white and brown chicken feathers off on my shirt and hair. This was when I spotted the three huge eucalyptus trees that stood across the street in front of the school. Two of the trees had smooth-skin on their trunks, but the other one had twisted-skin all about its bottom structure. I immediately liked the one with twisted-skin best. I could see that he was smiling like a huge, old white elephant as he watched the kids run past him. I nodded, "Good morning," to the huge tree, and he, of course, winked back at me just as *mi mamagrande* had always told me that trees will do when we address them with an open heart.

"Come on, *mijito*," said my mother, standing back up and closing our car door behind me. "I got most of the chicken feathers off of you. I can see that we're going to have to start closing our car windows at night, so the chickens don't try to nest on our car seats anymore."

I nodded. This was something that I really liked a lot about my mother and father. They were always thinking ahead so we didn't make the same mistake twice. Making the same mistake—which I used to do when I was little—could really be painful if you kept getting knocked on your ass by life.

I thanked my *mama* for cleaning me up, and she now walked me across the street. My *mama* looked so tall and beautiful in her new red high-heeled shoes and long, smooth, silver-gray dress. Her dress made a nice little swishing sound as she walked. She was carrying my lunch bag for me and holding my hand. The touch of her hand felt so warm and good that it made me feel good all the way up my arm. I loved *mi mama*. She, too, was my everything—just as my dad's *mama* had been his everything when he'd been small. My eyes darted everywhere. I'd never seen so many kids in all my life. And almost all of them were bigger than me and they spoke English and were laughing and having so much fun. I didn't see any kids that looked scared and were holding on to his mother's hand for dear life like I was doing. But I didn't care. I loved holding hands with *mi mama*.

But, then, getting to the three big eucalyptus trees across the street, I stopped. I didn't want to go any farther.

"No, *mama*," I said, pulling my mother down close to my face so I could whisper to her, "this is a bad school. I don't want to go."

"But how can you know that it's a bad school," she said. "You haven't even tried it yet."

"This tree, the old wrinkled one, he told me, *mama*." Trees had been speaking to me all of my life. Ever since *mi mamagrande* had taught me how to plant corn and listen to our vegetable garden. And old trees—not only liked to talk a lot, but they were also really worth listening to, my grandmother had explained to me, because they'd seen lots of life and so they were real smart.

"What did he tell you, *mijito?*" asked my mother.

"He told me that bad, awful things happen at this school."

"And did he, then, tell you to not to go to school?"

"No, he didn't tell me that, *mama*. He said that I'm going to have to be very careful and very strong at this school."

"You see, *mijito*, then this tree isn't telling you to not go to school here. He's just telling you that you must learn to be careful and strong, just like my mother's Crying Tree used to advise me back in *La Lluvia de Oro*, during the Revolution. So come on, *mijito*, you must be brave, and if anything really bad does happen to you, before I come back from the store, then you just run out and hug this tree, your friend, till I get back. Okay?"

"Okay," I said, feeling much better now. "And I promise not to pick my nose, *mama*, even if my *mocos* get dry and itch real bad."

"Good, and here, take my handkerchief," said my mother. "This way you can blow your nose like a gentleman, instead of picking your nose like a foolish *tontito*."

Hearing this word *tontito*, I laughed, because it meant dummy, but in a very affectionate way. Taking *mi mama*'s handkerchief, I felt so proud. Because I knew that *mi mamagrande* had especially hand-embroidered this handkerchief with the little red roses for my mother. Quickly, I put the handkerchief in the right back pocket of my Levi's, and we now went past the three huge old eucalyptus trees, through the wire-mesh gate that was way taller than me, and up towards the buildings of the school itself.

Inside the fence, kids were playing ball and running every which way. One boy came rushing by us so fast, chasing after a big white ball, that he bumped into me, almost knocked me down, and when he saw how I was holding on to my mother's hand, he laughed, calling me a "sissy" or something like that, but I didn't let go of my mother's hand. No, my dad had well explained to me that a real man didn't get offended if other men ridiculed him for staying close to the women of his *familia*. That a real *hombre* was proud of being close and loving with the women of his life.

After asking several people questions, my mother led me down a concrete walkway that didn't have any lumps of chicken shit on it like ours at home, towards the far building. Suddenly, without warning, a buzzer buzzed so loud that it scared the living hell out of me and I covered both of my ears with my hands.

Now kids were running fast in all directions and parents were waving goodbye and going out the wire-mesh gates to their cars. My mother and I were just about the only people left on the playgrounds. This was the first time I heard anyone else speaking Spanish besides my mother and me. This other Mexican mother, who had three little kids with her and seemed to be even more lost than my mother and me, asked my mother for help. My mother took her slip of paper, read it, then pointed towards the same building where we were headed.

I then saw that she had a girl who was probably just about my same age. But the girl looked taller and braver than me. Quickly, I guessed that this was the type of girl that my dad would say I should breed with—I mean, marry first, then breed. Because you see, ever since I could remember, my dad had been telling me that any breeder of fighting bulls or fighting cocks knew that when he finally found a good bull or cock, he didn't ask who the rooster or bull were. No, he asked who was the cow or the hen, because the cows who carried their young in their bellies, and the hens, who knew how to build their nests and sit on their eggs, had been given very special instincts by God. Women were the foundation of any home or tribe or nation, my dad always told me, so it was never too early for a boy to start studying girls, so he'd know how to choose the best wife. So I now watched this girl as she walked alongside her mother, as we all walked together to the far building. She was real pretty.

My mother knocked on the door. A tall woman opened the door, read over the slip of paper that my mother gave her, and then indicated to me to enter the classroom and go to the rear of the room. But the room smelled funny and all the kids in the front of the class were staring at me. I froze, refusing to let go of my mother's hand.

Then the tall woman, our teacher, read the other mother's form, and gestured her daughter to also go to the rear of the classroom, where all the other Mexican kids were located. And to my surprise, the tall, dark-haired girl kissed her mother, then turned and did as she was told.

Seeing this, I let go of my mother's hand. I was mystified by this girl. My God, she looked so brave, just walking right down the center aisle with all the eyes of the other students on her. Myself, I was ready to pee in my pants, I was so scared.

Suddenly it entered my mind that all these girls at school had also, probably, been told by their parents to start looking at us boys to see who would make a good husband. If this was true, then I was sure that no girl in her right mind would ever want me as a husband, the way I was behaving.

Quickly, I dried my eyes, made sure not to pick my nose—which was just itching to be picked—and I reached out with my right foot to take my first step towards that center aisle so I could go to my seat, too. But then, I don't know why, I just panicked, jerked my foot back, said to hell with what all these girls thought about me, and I grabbed hold of my *mama*'s leg this time—not just her hand—hanging on for dear life.

Some of the kids started laughing. I closed my eyes so I wouldn't have to see them and started praying. "God," I said, "please help me to be brave and not such a coward."

Then I started praying for Jesus, God's Son, who'd been so brave that He hadn't even cried out when they'd drove those big nails into his hands. Man, I would've screamed! I asked Jesus to come and help me to be brave like Him, and I was just beginning to feel better when suddenly the big, tall teacher came over and grabbed me, trying to pull me loose from my mother. I SHRIEKED at the top of my lungs. And I guess that I scared the living crap out of the teacher, because she now leaped back away from me, eyes huge with shock.

Regaining her composure, the teacher came at me again, and this time it became a tug-of-war with the teacher trying to pull me off of my mother's leg. But I was strong and I just wouldn't let go. Now the whole classroom was laughing, so finally the teacher and my mother both took me in hand and walked me down the center aisle and put me in my desk. But still, I wouldn't let go of my mother's leg. No, I kept crying and hugging my face into *mi mama*'s dress so that no one could see how big a coward I was.

"*Mama*," I said. "I want you to stay! Please, I don't want to go to school here! Something bad is going to happen!"

"Nothing bad is going to happen to you at school, *mijito*," said my mother. "Some of my happiest memories were going to school. So now let go of me. Don't you see how nice and quiet everyone else is behaving? You can do this, too, *mijito*. You're five years old. You're a big boy. Now, please, take your lunch, and let go of me."

Looking around, I could see that my mother was right. All these other kids were my age and they were already at their desks and not crying. My mother finally got me to let go of her, finger by finger, then she pushed me down, telling me to stay put. She gave me my brown paper lunch bag and I dropped my head, watching the back of my mother's beautiful red shoes go up the center aisle towards the front of the classroom, stop, talk to the teacher, who wore black shoes, then go out the door.

Seeing my mother's red shoes disappear, I almost leaped up screaming again, but then, the boy next to me said, "*Calmate*," in Spanish, "we're going to be okay, *mano*."

I turned and looked at this boy. My God, his Spanish sounded so soft and comforting, and he was the most darkly handsome boy that I'd ever seen. His eyes were as large and beautiful as a goat's eyes. Looking at him, I stopped crying. He was so calm and sure of himself.

I wiped my eyes, rubbed my nose clean with the back of my hand, and started to wipe it off on my Levi's, but then I remembered my mother's handkerchief, and so I brought it out and wiped my hand. This felt good. My *mamagrande* had done such a beautiful job when she'd hand-embroidered this handkerchief.

Looking out the window, I caught a glimpse of my mother's beautiful red hat as she walked out of the wire-mesh gate, and crossed the street to our car. I took a breath and put my face down into my *mama*'s handkerchief, smelled of her fragrance, felt better, and said another quick little prayer for Jesus to help me be brave, and made the sign of the cross, over myself which always felt very good to do, too.

"You don't have to be crying," continued saying the boy next to me in Spanish. "We're together back here, and we're going to be okay. Haven't you ever been away from your mother before?"

"Yes, but, then, I was always with *mi mamagrande* or my father or my sister and brother."

"Me, too," he said. "When I was little, but now we're all big, so we got to—"

This was when we heard a scream.

And it was a huge awful scream. "ENGLISH ONLY!" shouted our teacher at us as she came rushing down the aisle towards us Mexican kids in the back. There were about eight of us Mexican kids and three Black kids in the back of the classroom. All the other students were White and they were up in the front part of the room. "You two will not be whispering back and forth between each other, telling secrets in my classroom! Do you understand me?"

Her face was filled with such anger that I stopped crying. No, now I was so scared, I was ready to pee in my pants.

"Pee pee!" I said, standing up and holding my mother's hand-embroidered white handkerchief as tight as I could between my legs. All the kids started laughing.

"There will be no bathroom till recess!" she shouted. "Get back in your seat!" she added, grabbing me by the shoulders and shoving me back down in my seat. "You've caused enough trouble for one day! There will be no more special attention given to you!" Then she turned to the whole class. "Is that understood! We're here to learn, and that's what we are going to do: LEARN!"

I was sitting quietly with my eyes closed, hoping to God that no one noticed that I was beginning to pee. But my pee wouldn't stop, no matter how much I squeezed my legs together. And at first, my underwear and the handkerchief soaked up most of the pee, so I hoped to God that maybe this would be the end of it. But then, to my surprise, the pee just kept coming and coming and began working its way down the seat of my pants, feeling real warm as it formed a puddle of urine under me on the seat of my desk.

Then, incredibly, the pee still wouldn't stop and it continued coming in a steady flow, and now I could feel that it was going to start dripping off the sides of my seat. And if it did this, I just knew that everyone around me would start hearing the drip-dropping sound of my pee as it hit the floor like rain coming off the roof of a house.

I pulled my butt forward, through the warm puddle of urine on my seat, so that I could maybe get my urine to slide forward and run down the inside of my new Levi's. And thank God, it worked. I pulled my butt forward through the puddle of urine and I could now feel most of my pee running down the inside of my Levi's and going into my boots, warming my feet.

But I had no control of the smell, and soon the smell was beginning to cause the people closest to me to sniff the air and look towards me. This was when I saw the tall girl, who'd walked so bravely down the center aisle, turn and give me a look of pure disgust. I began to cry. I just couldn't help it. It was only my first day of school and I'd already failed. No girl in her right mind would now ever want me for a husband. I was a coward, and cowards were no good for nothing, ever. But then, I saw that one good thing about only having *Mexicanos* all around me, was that not one of them said a word, so the rest of the classroom never knew about my moment of terrible shame.

By the time recess was called, I didn't have to pee anymore. But still, I went to the bathroom, got in a stall, took off my Levi's, slipped off my underwear, and threw them and my mother's handkerchief into the toilet. But they wouldn't flush down no matter how much I tried. Then, horror of horrors, the toilet began to overflow.

"Oh, my God," I said, "why are You letting all these terrible things happen to me, *Papito?* Eh, is it that I did something wrong? Or are You just so busy doing big, important stuff that You forgot that we had a deal and You were going to stay close to me today."

I quickly put my wet Levi's back on, fished my mother's white handkerchief out of the toilet, and ran out of the restroom so that no one would know that I'd ruined the bathroom, with the toilet now overflowing all over the place.

Outside, I tried to squeeze all the urine and toilet water out of my mother's handkerchief. My heart was beat, beat, beating a million miles an hour. Then I noticed that the tall, old, wrinkled eucalyptus tree was right there by my side, just outside our playground and he was smiling at me. "Be brave," he said to me in a soft, kind tone of voice. "Be brave."

And saying this, his limbs took on wind and all his leaves began to sing and dance. I took a deep breath. I didn't know what to do, but he was being so kind and beautiful that I suddenly felt a lot less scared. I glanced around and saw that all the other Mexican kids of my classroom were at the far end of the playground.

"Go to them," said the huge old tree, winking at me. "And always remember what your mother told you, that school and life are much easier when we have friends. See, myself, I'm not alone. I have two good friends growing alongside of me."

And it was true; the old wrinkled tree had two smooth-barked trees growing right alongside of him. And they, too, were now dancing and smiling at me with the wind rustling through their leaves. I felt a whole lot better. "Thank you," I said to all three trees, and then I turned and started across the playground.

The darkly handsome boy's name was Ramón, and he was the one who was doing most of the talking when I came up. None of the other kids paid any attention to me, and who could blame them? I was the biggest crybaby of the lot.

"What are we going to do?" asked one boy.

"I don't know," said Ramón, "but one thing is for certain, they're treating us here like we're a bunch of *pendejos.*"

Pendejo was a very strong word in Spanish that meant stupid-ass, but really had even more science and *chile* to it than that. We all quickly agreed with Ramón's assessment of the situation and we were now voicing our own opinions and just beginning to feel better among ourselves when suddenly, out of nowhere, a huge muscular woman-teacher, who had a voice like a man, rushed in on us.

"NO SPANISH!" she bellowed. "You were all told that in your class-room! There will BE ONLY ENGLISH SPOKEN on the school grounds! Do you boys understand me!"

"Pee pee," I said under my breath, wondering if "pee pee" was Span-ish or English. And once more I squeezed my legs together real tight, hoping to God that I wouldn't pee again.

To my huge surprise, as the rest of us all went silent with fear, Ramón went right on talking in Spanish in a calm tone of voice, saying, "They're not our parents. They have no right to be yelling at us, espe-cially when we're out here by ourselves."

"I told you, NO SPANISH!" yelled the teacher, grabbing Ramón by his shoulders and shaking him.

"OYE! YOU'RE NOT MY MOTHER!" shouted Ramón at the huge teacher. "Let go of me! You have no right to be grabbing me!"

But she didn't let go. No, she now grabbed him by his hair, shaking him all the more. "NO SPANISH, I said!" she bellowed. "You hear me, no Spanish!"

"LA TUYA! VIEJA PINCHE MALA!" shouted Ramón.

"What did you say!" yelled the teacher. "Don't think I don't know your DIRTY SPIC WORDS!"

And she slapped him across the face, once, twice, three times, but still Ramón continued speaking in Spanish, telling us not to fear, that we were *Mexicanos,* and that we weren't their slaves!

I quit crying, just like that. My God, I couldn't believe it, this boy Ramón had to be the bravest human being I'd ever seen. And the huge teacher, she just kept right on slapping him until his face was covered with blood.

I dried my eyes and thought of the pictures of Our Lord God Jesus on the walls of our church when He'd carried the cross on His back to cavalry. Quickly, I made the sign of the cross over myself, and I was say-ing a quick prayer when the buzz, buzzed and she finally stopped hit-ting Ramón. And my God, Ramón was just a little five-year-old kid like the rest of us. She dragged him off, towards the bathroom.

"The rest of you dirty little spics get into your classroom RIGHT NOW, while I wash this little twirp's mouth out with soap!"

Another teacher, a man, came up but he didn't yell at us. He'd seen what had happened, and he very nicely escorted us back to our class-room. I'd been right when I'd told my mother that something bad was going to happen at this school. My mother had been gone less than an hour and already things were going so bad that it seemed to me like it was the end of the world.

That day, my mother never came for me until the end of school, and when I saw her, I got real mad and yelled at her. "*Mama,*" I shouted, "why didn't you come back to stay with me like you said that you would?"

"I did come back," she said, "but the school authorities told me that parents aren't allowed on the school grounds unless it's an emergency." Then she asked me if I was okay?

I shrugged. I didn't know what to say. Compared to Ramón, I guessed that I was okay. And at home that afternoon I tried to wash my own Levi's and my mother's handkerchief so no one would know my terrible shame.

Then at dinner, when my father asked me how my first day of school had gone, I didn't know what to say to him either, because I didn't want my dad finding out that his son was a crybaby coward, and that he'd been a fool to have ever told me that everyone loved Mexicans, because they didn't. Our teachers hated us Mexican kids!

So all I said was, "Okay, fine, *papa,*" and I said nothing more that day and I didn't say anything the next day either or the day after that. And each day, things at school got meaner and more terrifying than the day before. They were crucifying Ramón. They were really hitting him and hitting him, because he was the only kid among us who wouldn't break.

Finally, by the end of my first week of school, I was having nightmares almost every night and wetting myself in my sleep, too. And I'd never been a bed wetter before. I just couldn't help it. All night long in my dreams, tall, huge teachers kept chasing us Mexican kids, and they had big, sharp teeth like a dog's, and we kids had to keep running or we were going to be eaten alive if we spoke any Spanish. And sometimes, my God, Spanish would just slip out of our mouths.

School became a living hell, and at home I no longer wanted to listen to my father's stories around the dinner table about how great we, *los Mexicanos,* were. My dad was a STUPID FOOL! He had no idea what the HELL HE WAS TALKING ABOUT!

CHAPTER **five**

Looking back, I now remember that it was just about this same time when a red-headed kid with lots of freckles came to our school. And this kid spoke an English that was so much worse than all of us Mexican kids that we thought, Goodie, goodie! Somebody else besides us, *los Mexicanos,* is now going to get slapped across the face and called stupid!

But boy, were we wrong. Our teacher didn't hit this new boy, knocking the freckles off his face. No, she gave him compliments and always told him how smart he was. Finally we *vatos*-guys got together and went to our teacher to complain.

"Look," we said to her, "how come you don't hit him on the head and call him names like you call us? His English is way worse than ours. He's a foreigner."

"He isn't a foreigner," our teacher said to us. "You twirps are the foreigners. He's from Boston, and Boston is back east, and is a very important historical part of the United States."

Well, we didn't believe her. How could he be from the United States, talking as funny as he did. Like the word "car," he said *"ca-a,"* sounding like a crow talking to the wind. His name was Howard, but by the way he said it, none of us were able to figure out what his name was for the first few days. Still, I liked him, so one day I tackled him out on the playground and we started wrestling on the grass. He was strong and I was strong, so we were having ourselves a real fun time rolling around and hitting each other, when a bunch of my *vato amigos* came up and said, *"No te dejes! Pegale!* Don't let him get you! Hit him!"

Hearing this, Howard stopped wrestling and he stared at me with big eyes. "Are you Mex-eee-can?" he said.

I quickly translated what he'd said in Bostonian English into California English and said, "Yeah, sure, I'm *Mexicano.*"

Well, this boy, who'd just been wrestling with me with so much confidence and power, now let out a scream of sheer terror, scaring the living hell out of me, saying "YOU GOT A KNIFE!"

And I thought he said that he had a knife, so I turned tail and started running as fast as I could to get away from him. But then, as I ran, I figured out what he'd said and I turned around and saw that he was running in the opposite direction. I stopped and started after him, catching him at the end of the playground where he was all sweaty and scared and crying to beat hell.

"Look, Howard," I said, "I don't got a knife."

"Oh, yes, you do!" he yelled between sobbing cries.

"No, I don't," I said.

"Yes, you do!" he yelled again, his eyes flowing with tears. "My parents told me that Mexi-eee-cans always have knives!"

"They do? I didn't know that. I'll bring one tomorrow," I said.

The bell rang, and we walked back to the classrooms. Despite this, we really liked each other, and the next day I brought a little rusty pocketknife to school that I couldn't get open. At recess I showed it to Howard and asked him if he'd like to come to the ranch and go bow-and-arrow hunting with me or learn how to ride a bucking pig.

"I'd like to," he said, "but I can't even be around you anymore."

"Why not?" I asked.

"Because I had a talk with my parents last night," he said, "and they explained to me that I can't be around Mexicans because they're bad, dirty people and you can't trust them."

"Oh, I didn't know that," I said, feeling my whole chest tighten. "I'm sorry."

And saying this, I didn't know what to do, because I really liked Howard a lot, so I didn't want him getting in trouble for being around "bad, dirty people." I turned and walked away, feeling terrible.

I'd never known that Mexicans were bad, dirty people and you couldn't trust them. I'd just thought that we were stupid people, closer to the animals, and not as smart as White people, as the playground teacher kept explaining to us.

All the rest of the day, I hung around with my *vatos-amigos,* so I wouldn't offend anyone else, feeling really awful. And when my *amigos* asked me what was wrong, I didn't want to tell them what I'd just found out about us, so I kept still, with tears running down my face.

That afternoon, when I got home to our ranch, something happened to me that I would never forget for the rest of my life. I could now see very clearly that what Howard had told me about us, *los Mexicanos,* being bad, dirty people was absolutely true. Our horse corrals were full of manure and there were flies by the zillions in the milking barn, and our cars and truck and tractors were covered with dirt and mud.

That night at the dinner table, I felt like I was sitting down to eat with people I'd never really seen before. My mother, who I'd always thought was so beautiful, I could now clearly see that she wasn't. Her brown skin was the color of dirt and her dark eyes were too large, and her hair was black and her lips were too big. Also, she was chubby with large breasts and looked disgusting, the way she kept letting my baby sister Linda hang all over her, nursing all the time.

And my father, my God, he had a big head with curly black hair and a real thick neck like a bull, and he was so loud. He ate with his fingers, using his *tortilla* to scoop the food into his mouth, and he chewed with his mouth open, laughing and telling story after story, showing the food that he was chewing. I'd never realized it before, but my father wiped his hands off on the tablecloth, pulling more and more of the tablecloth towards himself as the meal progressed. We all had to move our chairs down the table towards him as we ate, if we wished to stay with our plates.

And my sister Tencha and brother Joseph, who I'd always thought were good-looking, too, I could now see that they were also chubby, dirty Mexicans just like my parents.

I remember well that I wanted to cry. I'd never known any of this about *mi familia.* We really were dirty, bad, ugly people, and liars, too, just as the playground teacher kept telling us. Because she was right;

we did lie to her when we'd deny that we hadn't been speaking Spanish on the playground. The truth was that we still spoke Spanish every chance we got. And why, because, simply, it felt good to hear in the sound of the language with which our mothers had rocked us to sleep when we'd been little.

I finally felt so sick, sitting at the table with these ugly, dirty, bad people, *mi familia,* that I got up and went to the bathroom to get away from them. Then in the bathroom, I'll never forget, I puked in the toilet, then stood on my little stool so I could wash out my mouth in the sink. And this was when I saw in the mirror, that oh, my Lord God, I, too, was Mexican and ugly! I had big teeth, a wide face with high cheekbones, and my skin was also a dirty brown color!

"Oh, my Lord God, *Papito!*" I screamed. "WHY DID YOU LET ME BE BORN A MEXICAN?"

That night I awoke screaming and my mother had to come to my room several times to help rock me back to sleep. And when she'd ask me what was the matter, what could I tell her? I didn't want my parents finding out that we were bad, dirty people and couldn't be trusted. I loved them. I really did. And so I didn't want to hurt them with all these awful truths that I was learning about us at school.

After all, the school was way bigger than our home, had a huge flag at the entrance of the school grounds, and so they knew what they were talking about; not my *papa* and *mama*—poor fools.

CHAPTER six

The very next day, I don't exactly know why, but I took two pocketknives to school. I guess I figured that, hell, since I was a no-good, dirty, bad Mexican anyway, I might as well carry two knives. Not just one. So I could then be the baddest, dirtiest Mexican of all. I'd found the first little knife in a drawer in the tractor shed. This second knife I took from my dad's personal toolbox. And this second one, a much larger knife, my dad used for castrating the pigs and calves on our ranch, and so it was razor-sharp and you had to be real careful opening it or you could cut yourself real bad.

At school I showed the *vatos*-guys my two knives, telling them how I'd been told that Mexicans always carried knives, so I was going to carry two.

"But that's not true," said Blackbird, one of the *vatos*, "in my family, we carry guns. Not knives!" This guy's nickname was Blackbird because he was the darkest *vato* among us.

"Yeah," said Screwdriver. He was the skinniest guy in our group. "My family has pistols and rifles, too, not just knives, so I say, we should also bring guns to school to protect ourselves from all of these *pinche* teachers!"

This made sense to all of us—why limit ourselves to just knives? We should bring guns to school, too. And so there we were talking, really enjoying ourselves, and getting to feel a whole lot better when suddenly, out of nowhere, there was that big, old muscular woman playground-teacher at our side, yelling:

"NO SPANISH!" she screamed. "ENGLISH ONLY!"

Then she saw the little knife that I was trying to open and she screamed out again. "OH, A KNIFE! I TOLD 'EM! I TOLD 'EM that this was what was going to happen! I was going to catch you little dirty spics with knives one day, and now I have!"

She was so happy that she was beaming. Quickly, she took the little knife from me that I hadn't been able to open, and she grabbed me by the ear, yanked me off the ground, and had me high-stepping it on my toes to the principal's office. The good knife, the sharp one that my dad used for castrating, Ramón had snatched from me when she'd first come up, and he'd hid it.

In the principal's office, the *gallo-gallina* teacher—as we'd nick-named her, the rooster-hen—yelled and yelled, saying how she'd risked her life, but still, she'd done her job and disarmed me just when I was getting into a knife fight with another dirty Mexican kid.

"A knife fight?" I said. "There was no fight! You're lying!"

"Are you calling me a liar?"

"Yes, but in English!" I said quickly. "All in English!" I added proudly.

Instead of my getting a reward for my brilliant ability to say all this in English, she now slapped me across the face so hard that I was knocked off my feet. Then the principal came around from behind his big desk, and I thought he was going to help me up and defend me from this crazy-*loca* woman, but he only started hitting on me, too, saying that he'd have no knife fights at his school.

Dodging their blows, I began screaming. "BUT I DID ENGLISH ONLY! Don't you see, *pendejos,* I did English only!" This was when I got a glimpse of my *vatos-amigos.* They were outside the principal's window standing on a bench, waving and laughing. That was the day that I was finally accepted by the *vatos* into their little club of *Posole* Town, the Mexican *barrio* in Oceanside, just up the hill from the pier and east of the high school. I'd finally turned out not to be such a little kindergarten baby-born-in-the-gravy, after all.

My parents were called to the school. They were told how I and a few other Mexican kids were the main troublemakers of the whole kindergarten, and that now we'd started a gang and that I'd brought a knife to school and gotten into a knife fight.

My mother was beside herself with shock. But my father, on the other hand, seeing the little rusty knife, just grinned and winked at me. "Does it open?" he asked the principal.

"What?" asked the principal.

"That little knife, does it open?"

"I don't know. I guess it does."

"Maybe you better check to see if it opens before you keep on talking and accusing people."

The principal tried opening the rusty little knife, but couldn't. "But this isn't the point," said the principal. "Your son brought a knife to school, that's the point!" And he went on and on about how bad we Mexicans were, until I thought that my dad was going to come out of his chair and knock the living shit out of the principal.

All the way home, my parents kept yelling at each other, trying to figure out what to do with me. I cried and cried.

"Lupe," my dad said once we were home, "be smart! Don't let yourself be taken in by that man's bullshit! There was no knife fight. Get it? That knife didn't even open. It will have to be soaked in oil for two or three days to loosen the rust before it can ever be opened."

"But, like he said, that's not the point, Salvador. What will our son be taking to school next, guns?"

"Yes," I said under my breath.

For the next few days all my parents did was argue about what to do with me. Finally, I couldn't take it anymore, so I decided that I was going to run away from home. I'd never meant to bring all this shame to my *familia*. There was just nothing else left for me to do but run away, was the way I saw it.

So Friday after school, I saddled up, got my Red Rider BB gun, and took off up the railroad track, going towards Vista. I figured that I'd go out past Bonsall, and learn to live off the land out by the Pala Indian Reservation where we had some relatives on my uncle Archie

Freeman's side of the family. Out there in the wilds of the Res—as our Indian relatives like to call the reservation—I'd live happy and free and be no trouble to anyone for the rest of my life.

I was five and a half years old and I'd already done-me nearly three damn months of schooling, so I figured that I'd just about learned-me all that I needed to know. "Shut up," "nap time," "keep still," "stop fidgeting," "no bathroom till recess,"—the only thing that they'd forgotten to tell us was when to fart, which was what some of us *vatos* had started doing after lunch in the back of the classroom. This drove our teacher crazy, especially when we cut loose a real-good, smelly one.

The Father Sun was just about going down into the ocean way behind me to the west when I came across two skinny, old trail-worn cowboys herding a herd of horses down the railroad tracks. They told me that they'd been driving this herd of stock all the way over from Arizona. They asked me if I knew of a place where they could hole up in some corrals for the night.

They were just about the two wildest-looking old cowboys that I'd ever seen. They had huge hats, almost looking like *sombreros,* big wild beards, and they were packing sidearms and smelled worse than a dozen road-killed skunks. They told me that when they'd started out at a border town near Tucson, Arizona, they'd had four other cowboys with them and well over a hundred and fifty head of wild mustangs. But within a couple of weeks of being on the trail, all four of the other good-old boys had quit on them, and so they'd had to sell off some of their stock, lost a few others, and now the two of them were down to just fifty-some head of horses after nearly six months of being on the trail. I told them that my family had a ranch and lots of corrals, but that I couldn't help them because I was . . . well, running away from home.

I'll never forget how they glanced at each other, grinned, but didn't laugh, then they looked over my horse and my equipment carefully. They saw that I had a blanket, my BB gun, a lariat, and my big red toothbrush was tied to the saddle horn.

"Well, it seems like you got everything," said the taller one. "So where are you headed?"

"East," I said.

"East?" he said, laughing. "Don't you know that cowboys don't ever go east? They go west!"

"Well, yeah, I know that," I said, "but our ranch ends at almost the ocean, so I can't go any further west unless I use a boat."

Hearing this, they started laughing so hard that I thought they'd die. Then the one who was doing all of the talking asked me if I could maybe postpone running away for a day or two and ride home and ask my dad if they could hole up in our corrals for the night.

"Yeah, sure, I could do that," I said, figuring that what the hell, it was a little too late in the day for me to run away anyway.

"Why—why—why are y'all running?" asked the other one, who hadn't said a word yet. He was shorter and smaller and had real big blue eyes shaped like a squirrel's.

I liked him. He seemed a lot more animal to me than human, which was good, of course, because my grandmother, Doña Guadalupe, had always explained to me that all humans were born with an animal-spirit to help guide them through life, and so the humans who realized this would always seem more animal than human, and this was wonderful. It kept us closer to God.

"Because . . . well, of school," I said with tears suddenly coming to my eyes. I couldn't talk anymore, I was so choked up.

"I-I-I done did-did-did the sa-same-same damn thing," said squirrel-eyes with his two big blue eyes getting even bigger. Then he started scratching himself all over, first at the back of the head, then he started in on his ribs, scratching wildly. I guessed that maybe school had rubbed him the wrong way, too.

"Me, too," said the bigger one who seemed to do all the talking. "School, I swear, it has ruined more wild-free-spirits than barbwire or a week of Sundays!"

Hearing this, I felt pretty damn good. And hell, I'd thought only Mexicans had it bad at school.

"Putting any young healthy boy or girl in a room and telling 'em to keep still all the time is as unnatural as putting a beaver on dry dock and telling him to forget his river-damming yearnings," continued the talker. "Hell, kids need to whip it up and be wild, for God's sake!"

Boy-oh-boy, these old guys were really talking my kind of language. Quickly, I turned my old horse Caroline around and started back for home on the railroad tracks. Never in all my life had I ever seen the looks of cowboys like these two old guys.

Getting home, I rode my horse past the corrals, right up to our house. *"Papa!"* I yelled, dismounting and running into the house through the back porch. "There's a whole herd of wild mustangs headed our way, and the two cowboys herding them want to know if they can hole up for the night on our place!"

"Where are they?"

"On the railroad tracks just below the cemetery out by El Camino road. They'll be here in a little while, *papa.*"

"I see," said my dad. "Then you already invited them?"

I turned all red. "Well, not really, but, well, kind of, I guess. Because I did tell them that we got lots of corrals."

"Okay, I'll back your hand this time, *mijito,*" said my father, "but in the future, I never want you telling people what we have till you've seen their hand first, especially with strangers. Eh, understand? You got to always keep your cards up close to your chest. That's a man's power. *Capiche?*"

"Yes, I *capiche, papa.*"

"Good."

I ran out the back door, got up on the porch railing, and remounted my horse Caroline and took off like a jackrabbit with a *coyote* on his ass. I needed to get back and help those two old boys bring in their herd of wild mustangs through that first narrow passage in the mud-flats. If a man didn't know those flats below our home, he could end up getting his livestock in a whole lot of trouble. And I knew all the country between us and El Camino like I knew the palms of my hands, especially now that we were getting hit on our hands so often at school that we kept needing to check our palms to see if anything had been broken.

"Yeah," I yelled, riding up fast to the herd, "my dad says you guys can hole up at our place for the night!"

Quickly, I took over the moving of the stock. Hell, I'd been riding horses since I was three years old and I'd moved a lot of cattle and horses in my day. It took no thinking for me to figure out who the lead mare was, and I quickly zeroed in on her and got her going down the safe trail across the mudflats. And the lead mare really knew her stuff, so working with her, I was able to keep the horses together as we came in towards our ranch.

My dad was at the corrals to meet us with a bottle of whiskey. I watched my dad's eyes look over the herd of horses real carefully as he handed the big bottle to the two old cowboys. The two old-timers each took a swig, wiping their mouths off with the backs of their hands. The taller one, the talker, introduced himself to my dad. He said his name was Jake Evans. Jake told my dad how the night before, they'd held up with their herd, just on the other side of Escondido—which lay about twenty-five miles east of us—and they'd been allowed to rest their stock up for three days.

Instantly, I could see where this was headed. Jake was now trying to pull a slick one over my dad and turn a one-night stay-over into three days. But my father didn't fall for it, and asked where they planned to be tomorrow night.

Jake saw it hadn't worked, so he then told my father that they planned on going all the way up to the Irvine ranch tomorrow, then work their way into Los Angeles by the end of the week, where they had a deal to sell their stock to a movie company that made Westerns just a few miles north of Hollywood. Jake then asked my dad if I could ride trail with them the rest of the way, since they'd seen how good a hand I was with horses.

I didn't know if they were pulling my leg or not, but I yelled, "Hell, ya! I'm ready! And I'll be saddled at daybreak!"

My dad and the two men laughed and passed the bottle of whiskey around one more time. Our conversation was all in Spanish. Our own ranch hands—who were *vaqueros* from Mexico—had gathered around and they, too, joined in the conversation. These two old *gringo* cowboys spoke their Spanish real good, and looked even more Mexican in their dress than some of our own cowhands.

My dad told the two trail-worn old boys that yes, they could keep their herd on our place for the night, and there would be no charge for the hay. They thanked him up and down and offered to pay him with one mustang, but he refused, saying that we had enough horses already, and they'd better hold on to what ever stock they had left. They looked mighty relieved and asked if they could shower, too.

"Of course," said my dad, "and after you've cleaned yourselves, you can come down to the main house and eat with us, if you'd like."

Both of the old boys turned all red. "No, *gracias, señor,*" said Jake. "We wouldn't rightly know how to behave in front of a *casa* with women and all. We'll just make us a fire out here on the ground and cook us up some beans."

"Look," said my dad, "wash up, come to the house, and if you don't feel comfortable, then just take your plates of *carne asada* and *salsa verde* to eat outside. My wife, Lupe, she makes the best damn *salsa* and *tortillas* you've ever tasted."

Hearing this, both cowboys' mouths started watering with wet, drooling saliva. "O-o-o-okay," said the nontalker. "Then that's how we'll—we'll—we'll do her!"

"Good," said my father, taking the bottle of whiskey. "I don't want you boys coming in all tanked up."

My father and I then walked home, carrying the big bottle of whiskey. *"Papa,"* I said all excited, "could I really ride with them tomorrow. I'm sure I could help." I wasn't that interested in running away anymore. No, now I figured that I'd just become a cowboy like these two old-timers.

"Mijito," he said, "it looks like fun herding horses, don't it?"

"Yeah!" I said.

"Well, it isn't. You just don't have any idea what those two poor bastards have gone through. Sure, in the old days, two real good horsemen could move a herd of horses from here to kingdom come and have no real *problemas.* All they'd have to do is watch out for rattlers, bears, *coyotes,* rustlers, and know where to find water and grass. But now, in modern times, with highways, towns, and barbwire everywhere, those two men got to be half crazy-*loco* to have brought a herd of horses all the way from Tucson, Arizona."

"But, *papa,* if I—"

"I've told you a thousand times, *mijito,*" he said, cutting me off, "there are no 'buts' or 'ifs' in life. If my aunt had balls, she'd be my uncle. And 'but' means that you didn't listen to a damn thing of what the other guy just said. No, *mijo,* you listen good, and understand that yeah, sure it looks like fun to you, and it could be fun back in the old days when a man was young, but neither one of these old cowboys are young anymore. They're old fools, *mijito,* trying to hold on to a way of life that doesn't exist anymore.

"They remind me of some of our own people back in *Los Altos de Jalisco* still trying to be *charros* to the very end. And I can admire them here in my heart for their spirit and horsemanship, but I'm not going to romance what they do here in my head, any more than I'll romance two people getting together and starting a family without having the means or guts to figure out how to make ends meet.

"Hell, it was almost dark when they pulled in here with that herd of horses. That's not right. A man, *mijito,* has got to have the *tanates,* get me, the balls to know how and when to change with the changing times."

I nodded. I could see that what my dad was saying was right. "Then it's once again like the little crow and the father crow story, eh, *papa?*"

"Exactly," said my dad, "each and every generation needs to add to the knowledge of the past generation, just like the little crow did to his father's."

This story of the little crow and the father crow, my dad had been telling me for as long as I could remember. How the father crow taught his little son-crow to be careful of two-legged people when they came close and bent over to pick up a rock. But then, the little son-crow thought even further ahead and told his dad that maybe they should fly off even before a two-legged bent over to pick up a rock, because maybe the tricky two-legged human already had a rock in his pocket.

"We can't just hold on to the past with wild, foolish hopes," said my dad, "and then get bitter and mad because things don't work out or stay the same. What would these two cowboys have done, *mijito,* if you hadn't come across them? And what would they have done if I hadn't decided to give them free hay? They got no money. You can see that by the shape that their stock is in."

I nodded. I could see that my father was right.

"Always remember," added my dad, "*dime con quién andas y te diré quien eres.*" Tell me who you walk with and I'll tell you who you are. "Your job is to go to school right now, *mijito,* and a good man does his job, no matter what."

"Really, *papa,* no matter what? Even school?"

"Yes, no matter what, even school. My *madre* didn't panic and die in the middle of the Revolution like my father and leave us kids to starve to death. No, she stayed with us kids, the three that she had left, and kept us alive no matter what. *Capiche?*"

I nodded.

"Good, then tomorrow, I'll send Tomas out with these two cow-boys to get them going," continued my dad. Tomas was our foreman. "Maybe the Marines will let them pass through Camp Pendleton. A lot of horses could get killed between here and Los Angeles if they aren't careful with traffic. Damnit, they're grown men, *mijo,* and you can't just be herding horses or cattle up the highways like in the old days any-more."

That evening, I didn't even recognize the two old cowboys when they came to the house all washed up and shaved. They were actually, maybe, even younger than my dad, and I'd thought that they were close to ninety-nine years old.

They decided not to take their plates outside to eat and sat right down with us at our big, old oak table. That night we heard story after story of how they'd had to fight off rustlers just on the other side of Phoenix, then they'd been hit by a hailstorm that had almost knocked them out of their saddles, then they'd had to go around on a long stretch of wild land to avoid the sandhill deserts at the California border so they could keep their herd from dying of thirst.

And now they, too, were wiping off their hands on the tablecloth like my dad. In fact, they seemed to just about copy everything that my dad did, figuring, I guess, that these were proper table manners. That night, I got to feeling a lot better about *mi familia*. I could see good in my *familia* once again, without all-that-stuff-from-the-school going off inside of my brain. School, in fact, now seemed so far away that I could hardly believe that I'd ever gone to that stupid place at all.

In the morning, I got up real early, planning to sneak off with the cowboys, whether my dad liked it or not. Hell, I wasn't ever going back to school again. I wanted to live my whole life up on top of a horse where I could run and leap like Superman, himself!

The two old cowboys were surprised to see me saddled and ready to go when they rolled out of their bedrolls. They'd slept in the hay barn, which was smart, because I figured that's where I would have slept, too.

"Are you sure that your dad said that you can go with us," asked Jake.

"Oh, yeah, sure," I said, lying, and suddenly remembering that this was exactly what they were always telling us at school that Mexicans did: lie. "But we got to get going fast, so we can beat the Sunday traffic going to church," I added.

"Look," said Jake, "last night, after you went to bed, we told your dad how we'd come across you running away from home."

"You did?" I said.

"Yes, we did. It was the honest thing to do, son. And you should've seen the hurt look on your dad's face, because, you see, Mexican kids don't run away from home. White kids, *gringo* kids, like me and Luke, we're the ones who run from home, but Mexicans, they ain't never do that.

"And do you know why they don't?" added Jake in that same soft tone of voice in which he always spoke. "Because Mexicans are just naturally warm, loving, good people."

"Really? Mexicans are good people?" I said, tears coming to my eyes.

"You're damn right! Great people! The best! Hell, all the way out from Arizona to California, in every town, they'd be the first ones to open up their doors to us two old coots and share their *frijoles* with us, no matter how poor they were. And that's what we saw around your dinner table last night, too. Good, old-fashioned, honest family hospitality with lots of love," he added.

"Big Jake's a'right!" said the other cowboy, talking with his half wild-nervous stutter. "Last night, she-she-she were the most, most—" It looked like he was just about ready to burst open, trying to get the words out that he wanted to tell. "—loving, damn good-good home-made-cheese night a'me life!"

"Luke's right," said Jake. "People dream of *una familia* like yours. Big, old, long dinner table with plenty of homemade cheese, *carne asada*, *salsa*, and fresh lemonade. Son, you got the whole world here, you understand, so you don't let no damn school cause you to run off. You dig in, you see, keep your powder dry, and don't let no son-of-a-bitch teachers run you off your homestead, you hear?"

"I hear," I said, feeling ten feet tall.

And that morning I stood shoulder to shoulder on horseback with my father and brother as we waved goodbye to those two old cowboys as they rode off with their herd of horses, headed for Los Angeles, then to a place north of Hollywood. Maybe Mexicans really weren't bad people, after all. So, then, maybe, it was all right for me to love *mi familia*. I had tears in my eyes watching these two old guys ride off, herding their stock.

"You really want to go with them, don't you?" said my dad.

"Yeah," I said.

"Well, then, go on," said our dad. "You and your brother go with them and Tomas as far as Camp Pendleton, then come home."

"My brother Joseph and I were off in a flash, slapping leather like real *CHARROS DE JALISCO!*

CHAPTER seven

I really don't remember too much more about the rest of my year in kindergarten. All I know was that, looking back, I could now see that those two old cowboys from Arizona were like a Godsend to me. After all the negative crap that had been pounded into my brain at school about Mexicans, no one in my family could have ever turned me around to believe that Mexicans were maybe okay people. But these cowboys had been able to do this. One, because they weren't Mexicans and were Anglos; and two, because they spoke Spanish as well as anyone in my family and they'd been armed and looked so tough, and yet, I'd also seen how much they'd admired, respected, and looked up to my dad and mom.

Then, a few weeks later, school was over and I was off for the summer. This also happened to be the year that my parents broke ground to build a new home for us on the ranch. I'll never forget as long as I live the big, old white-haired man who ran the construction site for us. His name was Englebretson and one day at lunchtime I saw him take his teeth out of his mouth and put them in a glass jar full of water and I took off screaming for home. He laughed and laughed and all that summer I watched them cut down trees and dynamite their trunks, blowing big pieces of stump some fifty yards away.

This was when I first learned that my own dad had worked as a professional dynamite man up in Montana with some Greeks in the copper mines, and that he still had his dynamite license. I loved what dynamite could do, so I paid close attention to everything my dad did to set a charge. Hell, with dynamite I could really do some big damage to anyone if they ever picked on me again, like they had in kindergarten.

One day, I'll never forget, I was with my brother Joseph when these two men drove up in a truck and put up tripods and set lines. I asked my brother if he knew what they were doing. Joseph explained to me that they were surveying. I asked what surveying meant and my brother explained to me that surveying was how you measured land, and that right now, these surveyors were giving instructions with their lines for the grading of our new home site. I still didn't understand, but then, when the big bulldozer arrived a couple of days later, suddenly everything started to make sense.

The big dozer climbed off the flat bedtruck and went right to work following the lines that the two surveyors had laid out. This was so fantastic! I could suddenly see what the word "grading" really meant. It meant that all the houses and streets of Oceanside had been surveyed first, then cut to grade, so that the rains would drain in the direction that you wanted them to go. This was fantastic!

I watched the bulldozer operator follow the survey lines with his great big steel blade, cutting at the soil for two days, and little by little, I began to realize that my parents were going to build the biggest damn house in the whole town! I was shocked!

"Are we rich?" I asked my brother.

"Yes," he said.

"We are? Then why do I always wear dirty, old work clothes?" I asked.

"Because we're ranchers," said my brother. "We're not city people."

"Oh," I said, "then it's okay for us to be dirty?"

"We aren't dirty," he said, laughing. "To be dirty means you never wash. We wash our clothes and take baths all the time. It's just that people that live on a ranch get dirt on themselves."

My eyes went big. I'd never thought of this. My brother was really smart. So this was also why so many of my Mexican friends came to school with dirty shoes and pants. Their parents worked in the fields. Boy-oh-boy, my brother was a genius!

This was also about the same time when I noticed that my brother Joseph, who was eight years older than me, could talk to the different workers on the construction site in English and in Spanish as easy as you please. He seemed to understand everything they were doing. He actually ended up helping the surveyors do some of their work. And one afternoon, I heard one of the surveyors tell the other surveyor, when my brother wasn't around, that Joseph was a real fast learner for being a Mexican.

I don't know why, but suddenly my heart was pounding. It was like, well, I was back at school, and once more, we, *los Mexicanos,* were being considered dumb, stupid, slow learners. I felt like getting a stick of my dad's dynamite and blowing these two surveyors' truck to smithereens! They would have never spoken like this about a *gringo.*

I began to dream about dynamite. I loved the stuff. Just the sound of the word itself, I loved! After all, I was big now. I'd completed kindergarten. And so this morning I got up real early so I could help my father with his dynamite work. We were now clearing an entrance to our new home. First we dug a little hole at the bottom of the huge tree stump that we wanted to blow up. Then we went to the tractor barn where my dad kept half a case of dynamite. He never liked to buy too much at a time, because dynamite could get very nasty, he explained to me, especially when it got old and started "sweating." After we got the half case from the cool dry place where we kept it deep inside of the barn, we went to the house where he kept the dynamite caps and cord. Because it was never a good idea, he'd told me, to keep the caps and cord stored right along with the dynamite itself.

Getting everything, my dad and I got in our old ranch truck and drove back over to the stump that we were to blow, which was about fifty-some yards away from our home site. This morning, my dad decided to use a few extra sticks on this huge, old stump.

Boy, it was a good thing that we used a long line and we got ourselves a good ways back, because when this big old stump blew, it blew to SMITHEREENS, then came falling down in big pieces just right in front of us, knocking the hell out of Englebretson's old truck. I mean, the roof of his big old pickup totally caved in and his windshield burst into pieces!

Englebretson came racing out of the construction site, yelling so wildly that his false teeth fell out of his mouth. I had to run and hide, I was laughing so hard.

"Calm down! Calm down!" my dad was telling the big, old building contractor. "Damnit, I know! I know! It's my fault! I should've had you move your truck! I just wasn't thinking!"

"You-you-you DAMN RIGHT, you weren't thinking?" yelled Englebretson the best he could. His teeth had fallen in horse manure and he had to wash them before he could put them back in his mouth. But this didn't stop him from trying his best to keep talking.

"Slow down! Take it easy," my dad kept saying. "I agree with you. A real professional don't have no accidents. Every man I ever saw killed in the mines, it was because he was in a hurry and wasn't thinking. There can be no rushing with dynamite. A good *hombre a las todas* is always thinking ahead and doesn't make stupid mistakes like I just did."

I was shocked. I'd never heard my father tell himself off like this before. But I could also see that he was right to do this, because we really should have had Englebretson move his truck, and also, by my dad saying all this, the old contractor was finally beginning to calm down.

Then I couldn't believe it, that very same afternoon, I got to go with my dad and old man Englebretson to downtown Oceanside, and I saw my dad pay cash to Ben Weseloh for a brand-new Chevy truck. Englebretson couldn't stop thanking my father enough. He'd never expected a new truck. He promised to make our home the best and strongest house that he'd ever built. He'd put 2 x 6s where the plans only called for 2 x 2s. He'd put extra cement in the foundation. He told my dad that he'd never had a brand-new truck in his life, and that this was wonderful!

Later, I heard my brother ask our father why he'd been so generous. "A man can never be too generous," said our dad, "when he's generous to a good, hardworking honest *hombre,* because that man will then break his back to do all he can for you. But . . . you be generous to a relative or a lazy, no-good worker, and they then think you're a fool, lose respect for you, and start thinking you owe them something, especially always-looking-for-a-shortcut people."

I watched my brother nod up and down as he chewed over our dad's words. "I think that you're right, *papa*," said my brother Joseph, "and also the cost of that truck is probably small compared to what Englebretson can now save you on building our home."

"Exactly!" said our dad with one of the biggest smiles I'd ever seen him give. "You got it *a lo chingón, mijito!* Generosity is a good investment when you know who to be generous with, and also, who not to be generous with."

"And how do we know who and who not to be generous with?" I asked.

Both my dad and brother turned and looked at me, and I could see that they were very happy that I'd been paying attention.

"That's the million dollar question," said my father. "And to learn how to do this, you watch, *mijito,* think, figure, and smell. Then, also, you don't get bitter and lose hope on people when you make mistakes. Because you can bet your boots that I've made my share of mistakes in the past on this one, and that's okay. This is how we learn, by making mistakes. And big ones, too!" he added.

I nodded. I was glad to hear this about making mistakes. "And big ones, too." I could now once more see that my father was real smart. He'd turned his terrible dynamite accident around, making it into a very good thing. Boy, had I been dumb to let my teachers at school ever convince me that my dad was a fool.

My brother Joseph started hanging out with Englebretson's son Chuck and this other Anglo boy named George. Both of these boys had horses and were just about my brother's same age. I watched these two guys change my brother's name from *José* or Joseph, to Joe, and my brother say nothing about it, because I could see that he really wanted to fit in with his new friends. After all, we didn't live in the *barrio* of Carlsbad surrounded by Mexicans anymore. No we were living in South Oceanside where there were mostly Anglos, except for our family on the ranch. And also, the year before my brother had started going to the Army Navy Academy in Carlsbad, and there were no Mexicans at that school, as far as I knew.

This morning, Chuck and George and my brother were dragging huge, thick, old railroad ties by horseback from the valley below us up the hill to the *adobe* wall park across the way from *Pozole* Town. They were helping to build an arena for a *rodeo* that was going to be held after the Fourth of July parade. And I wanted to help them drag the railroad ties with my own horse Caroline, but they told me that I was too little and my mare was too small and old.

Well yeah, maybe I was little and my mare was small and old, but I was sure "thinking ahead" enough to see this other accident coming way before it happened. They should never have tried to just keep yanking and pulling those damn railroad ties across the tracks like their leader Max Tinch was doing.

And I tried to tell my brother that this wasn't safe, and a good man *a las todas* was always thinking ahead, like our dad had told us, when he'd had that accident with the dynamite, but my brother didn't want to be left behind by his friends. So I watched him just keep trying to muscle tie after tie over the rails as fast as he could and *wham*—this time, the old tie got caught on the rail and my brother's old lariat stretched out all it could. Then the tie suddenly came flying off of the rail and hit my brother's right stirrup, mashing his foot.

The pain was so bad that my brother *José* turned white. I had to ride home as fast as I could to get someone to bring out the truck to take "Joe" to the hospital. I watched my brother's two cowboy friends laugh and laugh, saying that a real cowboy had to learn, one way or another. I almost pulled out my Red Rider BB gun to shoot Chuck and George. "You damn fools!" I felt like yelling. "Being stupid and in a big hurry has nothing to do with being a cowboy *a las todas, pendejos!*"

For over a week, my brother was laid up in bed and I ended up dragging over a dozen ties up the hill to the *adobe* park wall, but I first made myself a little ramp to get my ties over the rails.

That Fourth of July, eight of us from our *rancho grande* rode in the Oceanside Parade. My father rode his mare, Lady, a big Morgan Horse, my brother rode *Lasote,* a sorrel stallion, and I rode my little old bay mare Caroline. Two family friends and three of our ranch hands rode our horses with us. We were a big hit, the Villaseñor outfit, all dressed up like *Charros de Jalisco,* then afterward we all went to the *adobe* wall park to do the *rodeo.*

We, the little kids under seven, got to ride first and I was put on a squealing pig and rode that pig so well that they decided to put me in with the older kids to ride calves. I didn't want to do this, because I was still little and I could get my head kicked in, but Chuck and George and my brother—who was now walking about on crutches when he wasn't on a horse—kept telling me that a cowboy had to be tough.

"Tough, okay!" I said. "But not stupid!" I added, remembering what had happened to my brother when he'd been pulling those railroad ties.

But they wouldn't listen to me, so my big brother and his teenage friends put me in with the big ten-year-old kids and I rode a bucking calf clear across the whole arena, not getting bucked off once. But then I got thrown against the far fence and went facefirst into a pie of fresh cow shit. Then when I lifted my face out of the pie, I got kicked in the back of the head.

I was crying and crying and telling them that it had all happened just as I'd seen that it would happen inside my brain! But my brother and his two friends just kept laughing and wiped the shit off my face and told me that I'd won the event against kids almost twice my age, so I shouldn't be crying—I should be happy.

"Happy, hell!" I said. I was pissed and hurt and still crying. "Cow shit tastes awful! I'm never going to ride in a *rodeo* again!"

They told me I'd never be a real cowboy with that attitude.

"Good!" I said. "I don't want to be a real cowboy if you got to eat cow shit and get your head kicked!"

"You better tell your little brother," George said to my brother Joseph, "that if he keeps talking like this, people are going to think that he's just a little girl."

"Good!" I yelled. "Let people think I'm a girl! Girls are ten times tougher and smarter than boys!"

"Who the hell ever told you that one?" said Chuck, laughing, too.

"Tell him, *José*," I said to my brother. "Our dad told us that he saw our Indian grandmother rise up in the middle of the Mexican Revolution and save our whole *familia!* That she didn't fall apart and start drinking and die like our grandfather. Women are tougher and smarter than men, and they got to be! Because all life comes from between their legs! That's why the most important thing any man can do in all his life is to know how to choose the right woman for his wife!" I added.

But I could see that my brother was embarrassed with me, and George and Chuck thought that I was the funniest thing on wheels. Then when it was their turn to ride, they didn't do so well, and this wasn't funny to them, but it was sure funny to me. Both Chuck and George got their asses bucked off right at the start. My brother didn't ride in the *rodeo* because of his hurt leg.

It was from this day on that I began to notice a real difference between our *vaqueros* on the ranch from Mexico and the *gringo* cowboys. The American cowboys always seemed so ready to act rough and tough, wanting to "break" the horse, cow, or goat or anything else. Where, on the other hand, our *vaqueros*—who used the word *"amanzar,"* meaning to make "tame," for dealing with horses—had a whole different attitude towards everything.

To "break" a horse, for the cowboys, actually, really meant to take a green, untrained horse and rope him, knock him down, saddle him while he fought to get loose, then mount him as he got up on all four legs, and ride the living hell out of the horse until you tired him out, taught him who was boss, and "broke" his spirit.

To *"amanzar"* a horse, on the other hand, was a whole other approach that took weeks of grooming, petting, and leading the green horse around in the afternoon with a couple of well-trained horses. Then, after about a month, you began to put a saddle on the horse and tie him up in shade in the afternoon for a couple of hours until, finally, the saddle felt like just a natural part of him. Then, and only then, did a person finally mount the horse, petting and sweet-talking him the whole time, and once more the green horse was taken on a walk between two well-trained horses.

"You see," my dad told me, "for our *vaqueros* from Mexico, the whole idea is to make friends with the horse and for that animal to never buck. Not once. Sure, it takes longer, but you be patient and easy with a horse and he'll, then, end up trusting you and accepting you as his best *amigo*. Horse training is a courtship of *amor, mijito*. No different than bringing flowers to a girl and serenading her with music and dance. *Capiche?* This is how you show respect for the animal's heart and soul, recognizing that God is in every horse, *burro,* goat, pig, cow, and plant and rock."

I nodded. This, I understood. Then, also, on that Fourth of July, I found out that our *vaqueros* could ride bucking *broncos* with the best of the *gringo* cowboys if they wanted to. In fact, our top hand *Nicolás,* a real tall, lanky guy from *Zacatecas,* took first place at the *rodeo.*

It was the greatest learning summer of my whole life, but then came the fall, and I was told that I'd have to go back to school again.

"NO WAY, *JOSÉ!*" I screamed, because I now knew that at school they were trying to "break" us, not *"amanzar"* us. "I get a whole year off!"

"Who told you that?" asked my mother. "You only get off for the summer, *mijito*. Come on, we need to buy you some new clothes again."

But I was so angry with the thought of having to go back to school, that this year I refused to go to Penney's—my favorite store—to get new shirts and pants like I had the year before.

On my first day of school, I wore my old ranch working clothes, but I didn't care. It felt to me like I was returning to prison. My whole stomach hurt. To my surprise, this year they didn't scream at us right off the bat, because, I guess, most of us Mexican kids were now speaking quite a bit of English. Even Ramón. But he was a changed kid. There was a darkness in his eyes like a horse that had just been beaten one too many times.

This year we were given books to hold in our own hands and we were put in little circles to take turns reading. But first we were tested orally to see if we'd finished memorizing our alphabet over the summer. Hell, I hadn't even thought about school all summer, much less tried to memorize anything.

Ramón and I and a couple of other *vatos* were put in a corner so we could finish learning our alphabet. It was really embarrassing with the other kids—who all knew their alphabet—watching us. Finally, Ramón refused to be tested anymore, and I couldn't blame him, the way the other kids kept laughing at us. This was when it got so bad for us—the slow learners—that I watched Ramón close up as tight as a cow refusing to give milk.

But myself, being a coward, I was finally able to learn most of the alphabet except for the *c* and *z* which seemed to sound the same to me. And also, the *d* and the *b*, which gave me trouble because they looked so much alike. But still, I did well enough, so they finally put me in with the circle-reading-kids again, and I thought that I was doing pretty well until they showed us how to break the words down into groups of letters, saying that when these-letter-groups were put together, they made up the different parts of many words. Here, I got all mixed up.

Hell, it was just easier for me to memorize every new word that we were taught, and so I was doing pretty well doing it this way for the better part of the first grade, but then, when the sentences began to get longer than "Sally sees Spot" and "Spot sees Sally," I began to get behind on my memorizing.

I began to suspect that maybe, just maybe, something was wrong with me. I started developing a whole secret life in school that no one knew anything about except Ramón, who quickly figured out what it was that I was doing. And what I was doing was simply "thinking ahead" and figuring out who was going to be called on to read next, so that I could make sure to be gone to the bathroom or change my seat so I could act like I'd already had my turn at reading.

Sometimes this would be real hard to do, no matter how much I'd think and plan ahead, so when I did get called on to read, I'd lie and swear up and down that I'd already been called on two days before.

This was really awful! I didn't want to become a no-good lying Mexican, but every day I couldn't see any other way of getting around being called on to read. Life at school became a living nightmare, because now I could see very clearly . . . that I was becoming what everybody told me about Mexicans; they were stupid-liars and sneaky and couldn't be trusted, and why? Because we were no-good people.

But to not lie and have to get up and try to read in front of everyone and not be able to see the difference between "spot" and "stop," or "do" and "to" and have the teacher look at me like I was stupid and have the other kids laugh at me, was a fate worse than death. So I chose to lie, and change my seating so as to get out of reading.

Finally, the other kids caught on to what I was doing, too, and they told the teacher that I was a liar, that I kept changing chairs, and that I hadn't read out loud in class in weeks.

The teacher asked me, right there in front of the other kids, why was I doing this—lying and changing chairs. Did she really think that I was going to say in front of everyone . . . that I did it because I was stupid and couldn't read? So I just sat there with my head down like a "broken" horse, hoping to God I could disappear.

It was a kid named Fred, who wasn't a very good reader himself, who then broke out laughing and said, "He does all this because he's a stupid Mexican!"

All the kids burst out laughing and the teacher told Fred to not talk like this, and asked everyone to stop laughing at me. But after she did this, she just turned around and told me to go to the back of the classroom with Ramón and the other slow-learning Mexicans. And we weren't all *Mexicanos*. There were two other guys in our slow-learning section who were local Indians. The two local Indian boys were brought in every day to our school from clear out by the San Luis Rey Mission.

I put my book down and got up, and it was the longest walk of my life to go down the aisle between the desks to the back of the classroom. Everyone's eyes followed me. Some kids snickered, other kids laughed. And the teacher told them again to stop it, but if she really hadn't wanted them to do this to me, she could have taken me aside and spoken to me quietly and then told me to change my seating after recess.

Shit, on our *rancho grande* we treated our livestock better than they treated us Mexican kids at school. On the ranch, we never took a steer from the herd straight to slaughter. No, we first penned him up by himself, fed him real good for about a month, befriended him on a personal basis, then early one morning, took him down the road to the tractor shed, gave him grain, kept him calm, then killed him so quickly that he never knew what hit him. And this we did far away from all the other stock so they wouldn't catch the scent of blood and go crazy. I could feel that some of my classmates had gotten the scent of blood by the time I got back in the slow-learners' section.

And at recess, boy, did they come in on me and the other slow learners, like chickens on a crippled chicken, pecking at us with the most vicious, insulting words that they could find to hurt us. And hell, some of these kids—who were calling us names—they, too, were Mexicans like us, but because these kids had somehow managed to learn their alphabet and were reading, they now also felt superior to us.

I'd really tried so hard to memorize the alphabet, but I still had trouble learning my right from my left. After that public shaming by my teacher, school now went from being terrifying to becoming a living hell.

I began to get headaches every morning when my mother drove me to school, and I'd see my fellow classmates snicker at me even before I'd gotten out of our car. By the end of the first grade, I was no longer a very happy kid, not even at lunchtime when the other *vatos* and me got together to eat our *burritos*. We could see that all the other kids ate baloney sandwiches on white bread. It just seemed like there was *nada*-nothing that we Mexican kids could do that was right.

Ramón, on the other hand, was doing pretty damn well. Because, you see, by refusing to allow himself to be tested anymore, he had, in my estimation, cut right to the core of the situation and saved his ass and even his dignity. Myself, I'd become *un buey*, an oxen, a bull who'd been castrated and put to a plow.

Still, almost every day they'd find some excuse to send Ramón to the principal's office and he'd get hit or put in a corner, but this was a small price to pay—I was beginning to see—compared to the price I was paying by still trying to go along with what was expected of us.

And at home, I kept praying to God to please help me become brave like His Holy Son, Jesus, but it just didn't happen. I was a coward, a weakling. Just a no-good-for-nothing!

I began to wet my bed almost every night, and my parents kept asking me if something was wrong. But how could I tell *mi papa* or *mi mama*, that I'd found out that I wasn't just "a dirty, no-good, lying Mexican," but also "a stupid, slow learner" and a coward to boot. So I said nothing and my mother bought a rubber cover for my bed, as I mentioned before, and every morning, the first thing I'd do when I'd wake up was check to see if I'd wet my bed. And if I had, I'd quickly strip my bed and open up the window to get the stink out of my room. No one but my mother knew about my terrible secret. I'd sworn my *mama* to secrecy, asking her to please never, never tell my father or brother or sister.

I began asking my mother to please not give me anymore *burritos* with *chorizo* and egg to take to school for lunch. I wanted her to give me a baloney sandwich on white bread with a doughnut or Twinkie, like all the other kids. But my mother kept insisting that *burritos* were better for me. Finally, one day my mother surprised me and sent me to school with a baloney sandwich and a doughnut and a Twinkie in my paper bag.

I was so happy! At lunchtime, I'd sit as close as I could to the Anglo kids and I'd take out my baloney sandwich real carefully so they could see it. Then I'd eat it real slow so they could see that I was getting better, smarter, and learning how to do things right.

But then, horror of horrors, at lunchtime, I couldn't find my lunch bag. I looked everywhere. I ran this way and that way, trying to remember where I'd put it. Then, out of the blue, I kind of remembered, like way in the back of my head, that I'd left it on the bus. Because, you see, my mother was no longer driving me to school. I was now catching a bus. I went and sat down with my *vatos-amigos* and I was so hungry and feeling so lost that I swore to God that never again would I complain about my *burritos* if He just gave me something to eat.

It was at this moment that Ramón got up, stretched, and said that he didn't want to finish his *burrito,* and he offered it to me, then walked off, going to the bathroom. I took his half-uneaten *burrito* of egg with potatoes and a little *chorizo,* and I swear it was the best-tasting *burrito* I'd ever eaten in all my life.

That night, when my mother put me to bed, I told her what had happened, that I'd lost my lunch and I'd been so hungry I was going crazy. But then I'd prayed, and out of the blue, Ramón, my friend, had given me part of his *burrito* and it had tasted like the best food in all the world.

She laughed and told me that this was wonderful. "See," she said, "with friends, life is always easier. And with prayer, miracles do happen."

Then she had me say my prayers and thank *Papito Dios,* the Virgin Mary, and Jesus for the kindness that Ramón had shown me. Then *mi mama* began to sing me to sleep. Oh, my mother's singing to put me to sleep at night was quickly becoming my favorite time of each day.

"*Coo-coo-roocoo-cooo,* sings the turtledove," *mi mama* sang to me, soothing my forehead with her hand. "*Coo-coo-roocoo-coooo-cooo-coooo,* sings the turtledove," *mi mama* sang to me. "Close your eyes, *mijito,* says the turtledove, and your Guardian Angel will then appear to you in your dreams, and take you hand in hand like a bird, up, up to Heaven to be with *Papito Dios,* your Heavenly Father. Then in the morning you'll awake feeling all good and warm and soft and rested. Sleep, my little child, sleep. Go hand in hand back up to Heaven from where you came to visit with our Holy Family."

And each night, my mother would sing to me and massage my head and I'd fall asleep feeling so good. Then in the morning, it was true, I'd wake—having forgotten all about my terrors of school—and feel so good and warm and wonderful until I'd remember that I'd have to go to school again.

Later that year, I took Ramón aside and I told him that calfing season had come again and I needed him to give me back my father's castrating knife. "You see," I said to Ramón, "that's a very valuable knife. And the other day my dad was looking for it again and I had to pretend like I didn't know where it was."

"So why should this bother you?" asked Ramón in Spanish. "In school, this is all you do all day long, pretend to be a clown or stupid puppet for these *pinche* teachers."

I felt like Ramón had slapped me across the face. But I also knew that what he'd said was true. I was no longer *un Mexicano de los buenos*—a weed, a *yerba* that was so strong it would break concrete, reaching for God's sunlight. No, I'd become exactly what they'd told me I was; a stupid, dirty liar, and the biggest coward of all of us *vatos*.

"Look, Ramón," I said, beginning to feel tears come to my eyes. "I need my father's knife back."

"And what will you do if I don't give it back to you?" he said, not giving me an inch. "Rat on me, and tell your dad that you stole it and that I now have it?"

I just didn't know what to say or do. I finally shook my head. "No," I said. "I won't tell him."

"Well, then, good," he said. "I see that you still got at least one *tanate* left."

I wiped the tears from my eyes, and I knew that he was right, again; I was a *buey*. Hell, I'd already let them cut off one of my balls without even putting up a fight.

"Stop squirming," he said, "like you always do for these *pinche* teachers. Look," he added, "your family is rich. Everyone knows about the castle that your *papa*'s building. So grab hold of yourself, and hold tight to what you got left. Your father can buy himself a new knife. I need this one."

"But it's razor-sharp," I said.

He laughed. "Of course. What good would a knife be if it wasn't razor-sharp?" he said. "See you around," he added, turning around and walking off.

I took in a great big, deep breath, trying to stop crying, and I watched him walk away with that special walk of his. Ramón, he was just a little kid like the rest of us, but it was easy to see that he'd already become *un hombre*—just as my own father Juan Salvador Villaseñor had had to become a man at ten years old back in the days of the Mexican Revolution, in order to protect his mother and sisters.

I couldn't stop the tears running down my face. I, on the other hand, was a little crybaby coward, all confused and falling apart, but Ramón, like my father, they were *hombres a todo dar,* fighting cocks *de estaca,* willing to die before they allowed anyone or anything to cut off their balls.

Then I saw it. Oh, my Lord God, Ramón, he was like our very own Jesus Christ. I could now see this so clearly as he walked across the school ground. He had a glowing light all about him, because he, just like Jesus, was willing to carry the cross of crucifixion for all the rest of us lesser kids.

I dried my eyes, and made the sign of the cross over myself and then, strangely enough, I began to feel a little purring behind my left ear. And this purring, this little humming, this little vibrating behind my left ear, my grandmother had told me about, and she'd explained to me that people got this when they saw the magic glow of God's Sacred Light.

"Look out of the corner of your eyes," my grandmother had told me to do when we'd be watering her garden by the side of her little house in Carlsbad, "and you can sometimes see the Spirit of the corn, a gift given to our people by God. And you can see the Spirit of the squash and the string beans. All plants have Spirits. All animals have Spirits. It is only the two-legged human beings who have lost theirs. But you look close, and here and there, you will see the Spirit of a human glowing, too.

This is the Jesus in all of Us. This is the Sons and Daughters of *Papito Dios* inside all of Us."

I watched Ramón cross the playground, and the purring, the humming behind my left ear, began to travel from behind my left ear to the base of the back of my head, then all the way over to my right ear. And yes, now feeling the Sacred Circle of Life about my own head, I could see that Ramón was an Angel of God, just as *mi mamagrande* had told me that we all were, once we opened up our Hearts and Souls to *Papito*.

A few days later, it turned out just like Ramón had said that it would, and my father went out and bought himself another castrating knife. This one was black and from Germany and had two little men facing away from each other, or something like that, at the base of the blade. Hans, our good German friend, had ordered the knife from his brother in New York City, who sold German knives and other German kitchenware.

The knife that Ramón kept was dark brown and made in the United States, with the finest of American steel, and was so sharp that my father had always been able to shave the hair on his forearm with it before he'd spread the legs of the pig, calf, or goat to cut open the sack that housed the testicles.

I hoped to God that Ramón didn't hurt himself or someone with that terribly sharp knife.

CHAPTER **eight**

In the second grade, a bunch of us no-good Mexicans were told that we were being transferred to a temporary school out east of town, past the *adobe* wall park. But not all of us trouble-making Mexican kids got transferred. The tall, distinguished girl who'd walked down the aisle so bravely on our first day of kindergarten wasn't. She and a few of the other *Mexicanos* were kept at the regular school.

Instantly we, *los vatos,* could see what was happening. These Mexican kids who weren't being transferred were the ones who hardly ever talked to us other Mexican kids any longer, in Spanish or in English. No, they'd learned their English real good and now they mostly hung out with the Anglo kids. This was also when some of these smarter-English-speaking-*Mexicanos* began saying that they weren't really even Mexicans. They were Spanish, or even better, they were French.

And I could see why they'd done this, because now here at this other school, where we were almost all Mexican kids and just a few Blacks, things got so bad that I don't really remember very much, except that the janitor was a nice old man, and the only one who treated us with any respect or niceness. I mean, it got so awful at this school that now all I remember was that one boy, two grades above us, figured out that if we climbed up on the sinks in the boys' bathroom, we could then peek in on the girls going to the bathroom next door.

And we thought that this was really great, and we all took turns climbing up above the sinks until one boy just couldn't hold his laughter and the girls looked up and saw him looking down on them and they all SCREAMED BLOODY MURDER!

Then two girls came racing into our boys' bathroom, and they proceeded to beat the living smithereens out of the boy whom they'd caught looking down at them. Immediately, I remembered what my dad had told me about always being on the look-out for powerful, strong women, and so I figured that these were the type of girls that a smart guy should want to breed with—I mean, marry and then breed. So after the two girls knocked the hell out of that one boy, they turned on the rest of us, and I made the bad mistake of smiling.

"Stop smiling!" yelled the Mexican girl at me.

"Did any of you other guys look in to see us girls?" asked the other girl. This one was Black and looked as cute as a baby lamb, I thought.

"No!" said all the guys, lying at the same time. "We didn't see nothing!"

"You're lying, aren't you?" asked the first girl, *la Mexicana,* and she was very beautiful, too, especially her large, dark eyes.

"Yes," I said, still smiling, "we're all lying. And I want you two girls to know that I think you're the two most strong, beautiful—"

But I never got to finish my words, because it was now my turn to get the living hell knocked out of me. And damnit, I'd just been trying to make points with the two girls by being truthful. Besides, if the whole truth be known, we'd never got to see much of anything but the top of the their heads as they went in and out of their bathroom door. Hearing their toilets flush had really been the whole highlight.

Then, also, I don't exactly know why, but this was just about the same time when I began to draw stars. A teacher would be yelling at us, punishing one of my *vatos-amigos,* and I'd begin to draw a star. A five-pointed star. Then I'd color in the star with blue and a little bit of green. Soon this was all I did all day long, especially when things weren't going very well. I'd start drawing stars—five-pointed ones, then six-pointed stars, then I'd color them in with blue and green and little touches of red or yellow.

The blue and green made me feel real good inside. And the red and yellow seemed to somehow warm me deep inside, like the Father Sun. Soon, I really don't know how to explain any of it, but it felt as if I was creating a magic opening for myself—by drawing these stars—because then, when I'd colored a star, it would feel like I'd somehow magically jumped into it and be gone.

A couple of times, I hadn't even heard the teacher call on me, I'd been so far away, traveling through the Heavens in my star. This my *mamagrande* had also taught me—that people could star-travel, because that was what human beings really were, Walking Stars having come to Mother Earth to do *Papito Dios'* Holy Work, just like our big Brother Jesus.

When I finished the second grade, I got transferred to the brand-new school in South Oceanside on Cassidy Street. This school felt a lot safer for me, because it was so close to our *rancho grande* on California and Stewart Streets that I knew I could run home anytime if I wanted. Also, I noticed that there were hardly any other Mexican kids going to this school. Everyone was an Anglo, and I wondered why, but I didn't ask.

This year, I let my mother take me to Penney's to buy school clothes for me, and my first few days at this new school in South Oceanside were wonderful. I didn't know anyone and no one knew me, and so none of the kids knew I was stupid and made fun of me. This was really good, so I quickly made friends with a boy who lived up California Street from us, and his family owned the biggest bakery in the whole world in downtown Oceanside right on Mission Street, about five blocks up from the pier. Every day this kid brought cream puffs to school. I'd never had a cream puff before in all my life. At home, all we ever got to eat was the same old stuff: dirty ranch eggs straight from the chicken coop, lots of homemade *tortillas,* freshly made *salsa, carne asada,* great big avocados, tons of vegetables off our fields, weekly homemade cheese, and then, of course, all the fresh orange juice and lemonade you could drink.

We never got anything really good like cream puffs. The closest thing I'd ever had to a cream puff was *capirotada*—which was bread and cheese and *tortillas,* baked for hours with honey and raisins. But this was nothing like a cream puff—and hell, we only got *capirotada* once a year after our Catholic celebration of Lent. And then, of course, *buñelos,* during the Christmas season, were the only other sweet thing that we ever got. "Eat figs, eat oranges, if you want something sweet," my mother would tell us.

Well, everybody in the whole school wanted to be this kid's friend. His name was Whitakin, but I called him "What-A-King," which I figured was one hell of a good name. That year his parents bought him a brand-new Schwinn bicycle right after we'd started school. I loved his bike, so my parents bought me a genuine Schwinn, too. Now, every morning What-A-King and I would meet at the corner of California and Stewart Streets and he'd get off his bike and I'd get off of mine and we'd walk our bikes for a whole block as we each ate one of the two cream puffs that he'd brought in a little white bag. They'd taste so good, all soft and yummy, then we'd wipe off our mouths, lick our fingers, get back on our bikes, and continue to school. Once a month, What-A-King's parents would bring doughnuts and cupcakes for our whole class, but cream puffs, these he and I would get almost every morning.

One day, a new kid came to school who wore cowboy boots like me and he told us that he was from Texas, where he'd lived on a great big ranch with his grandparents before his father, who was a Marine, brought him and his mother out here to California. He and I quickly became friends and he was the one who taught me how to play marbles. Suddenly, just like that, everyone was playing marbles. I couldn't figure out how this happened. One day only a couple of kids played marbles on the whole school grounds, then this kid from Texas started playing and telling everyone that he was the best shooter in all of Texas, and now suddenly everybody and his uncle had marbles and they were all trying to beat this kid, but they couldn't. His name was Gus and he was really good and played for keepies, and so soon, he had the biggest bag full of marbles of anyone at the whole school, including even the fourth graders.

This was, also, the same time that I began to notice that things were becoming really different on our *rancho grande.* My sister Tencha was gone all week, attending a private all-girls Catholic school in Santa Ana, California, about two hours north of us. And my brother Joseph, as I said earlier, was now going to that military school in Carlsbad. Both my sister and brother now wore uniforms to school and I hardly ever got to see them except on weekends. But I remembered that my brother kept a bag of marbles under his bed, so one day after school I was going through his stuff, when he came in wearing his uniform and asked me what I was doing.

"I'm looking for your bag of marbles," I said. "I don't have any left. This guy Gus at school from Texas is beating us all and he plays for keepies."

"We had an Acuña kid in the *barrio de Carlos Malo* like that," said my brother Joseph. "He could beat the pants off anyone, even the older guys. But then he taught me how to do it, and I became a pretty good player, too, but never as good as Acuña."

"Could you teach me?" I asked.

"Sure," said my brother Joseph, and he changed clothes, got his bag of marbles, and took me out to the front yard of our new home that was being built across the field from our old ranch house, over on the knoll that overlooked our valley. There in the soft bulldozed dirt, he drew a circle with a stick and put twenty marbles in the middle of the circle.

"Now listen carefully," he said to me, "a lot of people are always trying to be first shooters, but being a first shooter is only good if you're so good—listen carefully, so, so good—that you can clean out the whole pot. Otherwise, it's best to be the second or third or even the fourth shooter."

"Why?" I asked. I'd always lagged my shooter trying to get closest to the line so I could be the first shooter.

"Because normally the first and second shooters only open up the pot of marbles, scattering them, and once that they're scattered, it's easier to shoot them out of the pot as singles, see?"

"But some guys come in with those big, old boulder marbles," I said, "and break up the pot with their first shot, then change from their boulder to a regular-size marble and shoot the hell out of the rest of the pot."

"Yes, you're right, that's what happens," said my brother. "And this is the genius of Acuña, he made up the rule of no-switchies."

"No-switchies?" I said. I'd never heard of this.

"Yes, that means that if you start with a boulder, you have to continue your whole game with that boulder."

"Damn, you can do that?" I asked.

"What? Make up rules? Sure, why not? All you have to do is have the other players agree to it."

"But how can you do this?"

"You vote on it?"

"You vote?"

"Yes."

"My God," I said, suddenly seeing a whole new world of possibilities, "then voting is damn good stuff!"

"You're right, voting is good stuff," said my brother, laughing. "You see, with the vote, then ordinary, everyday people can rise up together and have as much say as the powerful."

"Even me, who's not real smart, I can do this?"

"Sure. If you can talk, you can organize. Try it. Next time ask your friends to take a vote. Nobody really likes to see the pot broken up with a boulder, then see that shooter switch to a normal-size marble and shoot out most of the pot."

"Yeah, I don't," I said. "And Gus, he always starts with a boulder, then switches."

"Now, what I've told you so far, this still doesn't make you a winner," said my brother.

"It don't?"

"No, this is just the start, because now that everyone has agreed on using regular-size marbles, some people will want to bring in steelies."

"Steelies?" I'd never heard of these. I could see that my brother really knew a lot.

"Yes, these are the little steel ball-bearings that mechanics take off cars, and these steelies are worse than boulders, because a guy who can really shoot with a steelie—which was always difficult for me—will not only shoot out the whole pot, but that little steel marble will break pieces off your glass marbles and soon all everyone will have are bags full of chipped, broken marbles."

"You saw this happen?"

"Yes, that's why we also made the rule of no steelies."

"And you just voted on it like you did for the boulders? This voting business can really be good," I said excitedly.

"Yes, it is," said my brother. "That's why I'm going to become a lawyer, because it's only through the vote that we can help the poor workers in this country. Our father found this out in the mines when he worked with the Greeks in Montana. They'd vote on everything."

"Our dad knows about the vote?"

"He's the one who taught it to me," said my brother.

I was impressed. I'd never known that my dad knew all this and that my brother was this smart and such a fast learner, like those surveyors had said. I could now see that even though I was seven, I still had a lot to learn.

"Now, let me see you pick out the marble with which you're going to shoot," said my brother.

I quickly selected the prettiest marble I could find out of all his marbles.

"Go on, now shoot, so I can see how you do it."

I got down on all fours with my shooter in my right hand, ready to shoot at the pot of marbles in the center of the circle.

"Are you shooting at the whole pot?" he asked.

"Yes," I said.

"Okay, go on. Do like you normally do."

I shot, hitting the pot of marbles straight on, and nothing much happened. The pot just opened up a little, but not one marble went outside of the circle, and so my turn was over.

"Now watch me," said my big brother, and I watched him select the marble he was going to use as his shooter. And he didn't select the prettiest marble like I had done. No, instead, he took several, and weighed each one up and down in his hand, feeling each marble very carefully with his fingers and thumb.

"You see," he said, "I want the heaviest marble I can find that's still a normal-size marble, and also I don't want a real smooth-feeling marble, because then it's too easy for that marble to keep slipping out between my thumb and finger when I go to shoot."

I listened very carefully. None of this had ever entered my mind.

"I want an older-feeling marble," he said, "like a cowboy always wants an older, experienced horse when he goes out on trail, or a baseball pitcher who feels the stitches of a baseball real carefully before he throws."

"Wow!" I said. "This is so exciting!" A whole new world was opening up for me.

"Yes, learning can be exciting," said my brother. "And now watch, I don't shoot at the pot of marbles. I shoot at this lone marble that's over here all by itself, so I can knock it all the way out of the pot. That ensures me that I'll get another shot."

Well, lo and behold, my big brother Joseph cleaned out the whole pot. I mean, he shot out every damn marble in the circle. He was that great! No, he had to be better than great! He was the GREATEST SHOOTER I'd ever seen! I now know why he had the nickname of *Chavaboy,* Champion Boy!

"And now listen to me," he said. "I was never that good of a shooter until Mike Acuña taught me how to shoot, and I practiced for hours and hours."

"Then can you teach me how to shoot as good as you?" I asked all excited.

"You weren't listening. That's not what I said. I said that Mike Acuña taught me how to shoot, but then I added that I practiced for hours and hours all by myself."

"Oh, then, can you teach me how to do it?"

"Of course, that's the easy part. The practicing is what's tough. And while you practice, you don't play marbles at school."

"Why not?"

"Because it will interfere with you concentrating on getting better. For two weeks, all you do is practice at home alone, without telling anyone. And at school, you don't play. Instead, you watch this boy Gus and the others play. You study their every move, memorizing their different styles, learning a little here and there from every player, and you tell no one what you're doing. Understand?"

"Okay," I said. My heart was pounding. I was so excited, I could have popped. "I understand."

"Good, now first things first," continued my brother, "and so the first thing you learn is how to set the game up in your favor, or at least make it even. The second thing you learn is how to select your shooter, and now the third thing that you must learn to do is how to hold your shooter.

"You see, some guys like to hold their shooters, fist down, all knuckles to the dirt like this. Others like their hand rolled over, palm up. Then there are a couple of real fancy ways, up high with your fingers spread out. Myself, I learned to keep it simple so I could concentrate on the playing. So, if I'm first shooter, I take a double-hand stand, up high on my first shot, so I can shoot downward to break up the pot. But then, after my first shot, I go into the traditional, old-fashioned, palm-up, big-knuckle-anchored-in-the-dirt position. Are you getting all of this?"

"Yes," I said, nodding up and down. My God, and all I'd done was come into my brother's room to look through his stuff for his big bag of marbles. I'd never dreamed that there was all this "thinking" to playing marbles.

"Good," said my brother, "I'm glad that you're finding this interesting, because being able to not just 'hear' but really listen is what *papa* always says is the beginning of all learning. And to learn is to be able to make distinctions. Now, are you ready for the big one, the secret that will make you a winner?" he asked.

"Yes, I'm ready!" I said, feeling all excited.

His face changed and he put his hand on his forehead.

"What is it?" I asked, suddenly seeing that my brother wasn't looking very well.

"Oh, I don't know. I've just been a little tired lately," he said. "But now listen, this is the secret to winning, and this secret you must never tell anyone until you retire from the game, or you want to pass it on to your brother or sister—"

"Sister? Girls don't play marbles!" I said, laughing.

"In the *barrio* some girls did. Not too many, but the one or two who did, they were better than most of us guys."

"Really?"

"Yes. Now, listen closely, because this secret is actually done before you even start to play, and yet the whole outcome of your future is based on this secret. And *papa* says the same thing is true in poker, dice, business, in all games of life, there's a secret, and it's always very simple."

"Well, what is it?" I asked.

"Are you listening?" he said.

"Yes," I said, "I'm listening, damnit!"

"If you're the first shooter, you make sure that not all the marbles are put together real tight," he said.

"What? That's it?" I said.

"Yes," he said. "That's it."

"But—"

"*Mundo,* 'but' means you weren't listening," said my brother just as our father always did when we used the word "but." I could see that something was really hurting my brother. "Now, let's try again. If you're the first shooter, you make sure that when everyone puts their marble in the circle, two or three of the marbles are a little bit away from the rest of the pot. Then these loose ones are the ones that you shoot into, and from an angle. Because when you hit those loose marbles from an angle, they'll scatter away from the pot, and most often, one or two will go out of the circle. But if they're all in real tight, it's like hitting a brick wall for the first shooter."

"But *José,*" I said, "how will I ever know if I'm going to be the first shooter, since we always put our marbles in the pot before we lag our shooters to see who gets to go first?"

"Just change your procedure of doing this, too," he said. "Make a side bet of one marble each for whoever wins the lagging, so that the lagging will become a betting game on its own, and you can then say that you want to do it first."

"I can do that? Just tell people that we lag first, then we put in our marble afterward?"

"You can with the vote," he said, "and by being the one who made the suggestion of what to vote on."

"Really?"

"Yes. You see, once you get a vote going and people feel that you're a fair person, then they'll go along with almost all your other suggestions."

"They will? Oh, wow! Then the guy who brings up the idea on what to vote on is way ahead of the game of getting what he wants."

"Exactly."

"And *papa* taught you this?"

"Yes, and he learned a lot of this from—"

"His mother," I said. Our dad was always talking about his mother, Doña Margarita.

"Exactly, from his mother, and also from that man named Duel up in Montana."

Our father always spoke about Duel, too. "Then *papa* used to play marbles?" I said.

My brother laughed. "No, *papa* never played marbles," he said. "But one day when *papa* saw me looking sad in the *barrio,* he asked me what was the trouble, and I told him that I'd lost all my marbles. He asked me who had beat me out of them. I told him Mike Acuña. He told me to go and get Mike. I went and got Mike, thinking that *papa* was going to tell Mike to give me my marbles back, because he was way older than me, but *papa* surprised me. Instead, he asked Mike if he'd like to make fifty cents a day teaching me how to play marbles."

"Our dad did what? You mean that he paid someone good money to teach you how to play a kids' game?"

"*Papa* always explained to me that there is no such thing as a kids' game. That there are only games with which kids are learning the facts of life, but it's the parents that are so *tapados*—so blind and constipated that they can't see what these games are really all about.

"I'll never forget that day, Mike's eyes became larger than a rabbit's, because he'd thought that he was going to get in trouble, but instead, he was being offered a paying job. And fifty cents a day was a lot of money back then. 'It will take me a week,' said Mike. 'Make it two weeks,' said our father, 'and I'll pay you five dollars up front to get started, but you tell no one, understand, or I won't pay you the rest of the money. And you teach him here in the backyard, so I can keep an eye on both of you.' After those two weeks, I went from being one of the worst players in all the *barrio* to one of the best."

"I'll be damn!" I said.

"Don't you use that word again," said my brother. "You've said it, I think, three times since we've been talking."

"What word?"

"Damn."

"I have?"

"Yes, and you don't want to say that word towards anyone, and especially not towards yourself. Our *mamagrande* Doña Margarita, *papa*'s *mama*, would never allow us to use that word. Instead she'd have us bless ourselves and bless everyone else, too, explaining to us that this was our power, when we humans all lived in God's blessings."

"You knew her?" I asked.

"Of course, Tencha and I grew up with her."

"I never got to meet *papa*'s *mama*," I said.

"Yes, we all know that, because she died two years to the day, to the hour, before you were born, which means in some Indian ways of thinking, that you are her."

"Me? I'm our *mamagrande*?"

"Why not? We're all little pieces of somebody from our ancestry."

"You mean like pieces of dust from stars like *mama*'s *mama* used to tell me?"

"That, and also from our *antipasado*, our ancestry. Why do you think *papa* is so smart? He's lived life is many trials. He's his grandfather

Don Pio, his mother's father. Don't you always hear *papa* saying that blood knows blood and usually jumps one generation."

"Yes, but I never knew that he really meant what he was saying."

"*Mundo,* you start paying closer attention to our *mama* and *papa*. They didn't get this far in life because their eyes are closed."

I nodded. I'd never had a conversation like this before with my brother Joseph. I now wondered if his blood had "jumped," too, and he was then, in fact, our own uncle *José,* the Great, our *papa's* older brother who'd almost single-handedly saved all of *Los Altos de Jalisco* from the Mexican Revolution.

All that afternoon, I practiced my shooting, and I also tried to get better at selecting my shooter. Not for prettiness, but for weight and feel. Finally, I found that if I closed my eyes, I could feel better than if I kept my eyes open. And also, I found out that, when I finally found a good, heavy shooter that was a little bit rough and fit real good between my thumb and index finger, I could then put extra pressure on that marble without it slipping out of my hand. Suddenly, I was shooting way harder and straighter than I'd ever shot in all my life. My God, when I was finally called in for dinner, I didn't want to go inside. Learning was so much damn—I mean, blessed—fun when it made sense!

All that night, I dreamed of playing marbles. Something really wonderful had happened to me. It was like, before, I hadn't known enough so I could think or dream about playing marbles. But now that my brother had taught me all this stuff, I couldn't turn off my brain, especially when I went to sleep. It was like all night long I'd have that humming, that purring going on at the base of the back of my head, going back and forth, between my two ears.

In the morning, I awoke feeling like I had really slept in the arms of *Papito Dios* all night long. This day, for the first time in my life, I was all excited to go to school. But then, getting there, I quickly remembered that I wasn't supposed to play for two weeks. The next two weeks were the longest of my life, as I practiced and practiced at home every afternoon with my brother giving me pointers now and then. But he couldn't help me too much anymore. He'd made the junior varsity football team and he was now staying late at school and working out with the team.

And every night, I'd continue to have these wonderful dreams of playing great games of marbles, and I'd be really good. My daydreaming and my night dreaming had become one, just as my *mamagrande* had explained to me it had been for all people when we'd lived in the Garden. At school, I'd watch Gus and the other guys play, and now, way different than before, I could see things going on that I'd never seen before. I was making distinctions, just as my brother had explained to me I would do as I learned more and more. Gus, I could now see, wasn't just good at breaking up the pot with his big boulder; he was also, probably, the best shooter of the whole school, once he switched over to his regular-size marble.

My dreams of playing marbles continued, and then one night, my two dead grandmothers and my old dog Sam, who'd got run over by a car—saving my life—came to me in my sleep, and Sam licked my face and my two *mamagrandes* took me hand in hand and walked me across the Heavens. Then we stepped into a Star that was blue and green with a touch of red, and miraculously we were shooting across the greatest star-studded marble game ever played!

This was great! Instantly, I understood that playing marbles was no kids' game at all! Playing marbles was teaching me all about God's Garden, just as my grandmother had taught me when we'd planted a vegetable garden next to her little *casita*. Our *papa* had been absolutely right when he'd told my brother Joseph that there was no such thing as kids' games.

That morning I awoke, feeling that little purr-humming behind both of my ears, and I knew that all last night I'd truly slept in the Holy Arms of *Papito Dios*, just as *mi mama* told me I would do when she'd sing me to sleep each night—*coo-coo-roocoo*-ing me with the song of the turtledove.

The next day, I went to school with a strength and confidence I'd never felt before. And when I watched the guys playing marbles at recess, I could now see even more things than I'd never seen before. And yeah, sure, I knew that I wasn't supposed to play yet, but this didn't mean that I couldn't help to organize a vote. I mean, Gus was just beating the pants off everyone because of the great big boulder that he used for busting up the pot of marbles, and everyone assumed that this was the way it was and so we couldn't do anything about it.

I took a few guys aside and told them that I'd found out that switchies weren't allowed across the lagoon over in Carlsbad.

"So?" said one guy.

I could see that they hadn't understood what I'd said.

"No switchies," I said, "means that Gus can't change from his boulder to a regular-size marble after his first shot. That if he starts with a boulder, then he has to continue with a boulder, and no one is a very good shot with a boulder on singles, not even Gus."

Their eyes suddenly lit up. They loved it. Hell, we didn't even have to take a vote on it. They immediately just went over and told Gus that switchies weren't allowed in California. Well, I'd never said this, but it worked. Because no matter how mad Gus got about this, all the guys stuck together and said that they wouldn't play with him anymore if he continued switching from his big boulder to his regular shooter after his first shot. Finally, Gus saw that his goose was cooked, and so he calmed down and agreed to the new rule, but not until he'd cussed out all of us chicken-liver, sissy-baby Californians.

But still, even with this rule, Gus kept winning most of the marbles. He was just a damn good—I mean, blessed good—shooter. And at home, I kept practicing and practicing, and I could see that I was really getting pretty good. But now my brother wasn't helping me at all anymore. No, he was coming home after football practice and immediately going to bed, he was feeling so tired.

Finally, my two weeks were over and I could hardly sleep that night, I was so excited to get to school so I could start playing. "Dear God," I prayed that night, "please help me to play well tomorrow. And also help me to remember everything that my brother has taught me. Because, You see, God, my brother isn't feeling well enough to help me anymore.

Good night, *Papito.*" I added, "see You in the morning. And please help my brother *José* to get well."

That night I didn't dream of playing games of marbles. No, instead, I dreamed of all the things my brother had instructed me. "You're going to have to be very careful to not become a show-off or make fun of people who don't know how to play as well as you. . . . In fact, *papa* told me that Duel, his friend from Montana, explained to him, that if you get cocky or show off too much, people will turn against you and then you'll have no one to play with. So what do you do?" *José* then asked me.

"I let the guys win back some marbles, especially towards the end of the day," I said in my dream to my brother, as he'd instructed me to do, "so that then they'll go home feeling good and have the hopes of beating me tomorrow."

"Exactly," said my brother. "And you never brag about how good you are like Gus does, because a real king doesn't need to tell anyone that he's the king. In fact, a real king keeps his reign as much of a secret as he can. Good luck," added my brother, "and always remember that our father says that luck is a beautiful woman who needs constant courtship."

Waking up that morning, I was so excited to get to school so I could maybe pick up a game or two before school started, that I even forgot to wait for What-A-King, so we could eat our cream puffs together. I was into my second game of marbles, beating the pants off everyone, when What-A-King came up on his Schwinn bike. He was mad as hell and said that he'd waited for me at the corner of California and Stewart Streets for a long time. Then he'd gone to my house to see if I was sick, and our dog had almost bit him. He told me that he was never going to bring another cream puff for me as long as he lived.

I told him that I was sorry, but I couldn't talk right now because at that moment, I was winning. He left in a huff, and I continued playing. It was so exciting for me to win for a change that, boy-oh-boy, this tasted even better than a damn—I mean, blessed—cream puff!

All that week, I avoided playing against Gus, figuring that I had better do as my brother had told me to do, and that was to get quite a few wins under my belt before I played against the best kid on the block. On Friday, since I'd been winning every day, I figured that I was ready to go head to head with Gus. Also, all week he'd been making chicken sounds at me, calling me a sissy, and I was sick of it.

"Okay," I said to Gus on Friday morning as soon as we got to school, "I'll play you today, but only head to head, with no one else in the game with us." My brother had also suggested that I do it this way so I'd be able to keep my full concentration like I did at practice.

"But maybe Gus won't go for it," I'd said to my brother.

"Don't worry. Being a show-off, he'll jump at it," my brother had assured me.

"Now you're cooking!" said Gus excitedly, just as I'd been told that he would do. "*Mano a mano,* I like it! We'll all have us a regular Mexican standoff!"

Well, I liked the expression *mana a mano,* but I'd never liked the expression "Mexican standoff." "Look," I said, "just the two of us, and . . . and no switchies."

"You, too!" he yelled. "Hell, I'd thought you were a real cowboy! I didn't think you were a chicken-liver little girl like these other damn sissy Californians!"

I closed my eyes, breathed, then reopened them. "I am a real cowboy," I said. "That's why I'm a *vaquero.* And being a *vaquero* means—"

"I knows what a *vaquero* means," said Gus. "I wasn't born yesterday. It all means cowboy in Mexican."

"Yes. And the cowboys learned everything that they know from the *vaqueros.* The word '*rodeo*' itself comes from Mexico, and for you to call me a girl is good, because—"

"Yeah, I know!" he said. "I ain't an ignorant Californian! I'm a Texan! And we Texans all knows that our cowboy words and stuff originally came from the *vaqueros.* But then we Texans took over after the Alamo and improved on the whole thing so much that that's why we can now out-*rodeo* any damn Mexican *chile*-belly ever born!" he added with pride.

I felt like I'd been slapped across the face, because this wasn't true. I'd seen our top hand *Nicolás* out-ride all of the American cowboys last

Fourth of July once again. "No, not improved," I said. "Made mean in wanting to break the horse, instead of *amanzar* the horse, which means—"

"Who cares what it means!" said Gus. "Let's just stop jawing and see who's top dog even with your little-girl chicken-stuff of no-switchies."

I was so mad that I was trembling, so when we lagged, I wasn't that careful, and he easily won and got to be first shooter. But still, I knew enough about what I was doing that I made sure that all the marbles were in really tight together in the center, so he'd hit a brick wall with his first shot.

And it worked, just as I'd been taught that it would. Now it was my turn and I'd calmed myself down enough so I could shoot. I went after an easy lone marble like my brother had taught me to do. Nothing fancy. I didn't try to go after the two marbles that were real close together, so I could maybe knock out two at the same time.

No, I carefully went to work like my brother had taught me to do, marble by marble, concentrating completely on my shooting just as I'd done at practice, and before I'd even realized it—oh, my Lord God, I'd cleaned out the whole damn—I mean, BLESSED—POT!

I was even better than I'd thought. But Gus was far from giving up. And this time we didn't have to lag to see who went first. Now, since it was just the two of us playing, I got to go first. A whole bunch of kids gathered around to watch us play, even some girls. But I didn't let any of this distract me and I made sure that three little marbles were just a little tiny bit away from the rest of the pot. Then I acted like I was going to shoot from the other side of the circle like my brother had taught me to do, so when I finally came back around to take my angle shot on the three loose marbles, it would look like I'd just seen these loose ones by chance and hadn't deliberately set them up like this in the first place.

My God, it worked like a charm, and I cleaned out the whole pot again. By now, half of the school was watching. In fact, when the recess bell rang, not one of us ever heard it, we were all so excited. Our teachers finally had to come out to get us.

"That was just damn luck!" said Gus to me as we went back to our classroom. "Just wait till after school, I'll get you! I know you're not that good! Hell, I'm the one who taught you how to play."

I nodded, saying nothing. But boy, was I ever tempted to tell Gus that I'd been practicing at home for weeks and being coached by a player far better than him. I said nothing, not then, and not ever. This, I'd also been taught. "Keep your cards close to your chest," my father always told us.

That afternoon after school, Gus and I got together and we decided to do our shoot-out at the far end of the school yard under the big eucalyptus trees that grew along Stewart Street on the east side of our school. Here, we cleaned out a good-size piece of ground by dragging our shoes across it and wiping it all clean. Then we used fallen branches to sweep it smooth. About a dozen kids stayed after school to watch us play. Two were girls, and one was real pretty, but I was learning that pretty wasn't the best way for you to pick a girl or a marble. No, you closed your eyes and got a feel for the girl or marble. Closing my eyes, then reopening them, I could now see the second girl was real cute and had eyes as beautiful and kind-looking as a goat's, which, of course, was good, because this meant she was at peace with her animal spirit.

"All right," said Gus. He'd just finished drawing the circle. "Let's play!"

Looking at the circle, I knew that something was different, but I couldn't quite put my finger on it because everything looked so different out here underneath these huge eucalyptus trees. Then it hit me. Hell, Gus had drawn the circle way bigger than the circles that we normally used on the school grounds under the pepper trees, where we parked our bikes.

A chill went up and down my spine. I could now see that Gus had also had some adult coaching on how to play marbles. He'd just changed the whole game on me, and I could see why. He was bigger than most of us kids, and stronger, too, and so a larger pot would work in his favor.

"Pock, pock, pock," he said, making those chicken sounds at me when he saw how I was studying the size of the circle. "Come on. Let's just play. Stop being a chicken. I figure that you and I are the best damn shots in the whole third grade, and I already gave you no-switchies, so let's stop all this baby girl stuff and play like we do back in Texas. A big ass-old circle, and we put in fifty marbles each."

"Fifty!" I said, taken aback by this. "I don't think I even have that many." My brother had also told me to never bring all my marbles to school, only bring thirty or forty marbles at a time so no one would ever know how much I was winning.

"The guys who bring all their marbles to school," said my brother, "showing everyone how much they've won, are fools, and fools always end up out-fooling themselves."

"All right," said Gus, showing everyone his huge bag of marbles, "I'll just match whatever you got."

My heart was pounding, and I wanted to say no, and tell him that I'd only put in ten marbles like we normally did, but he started in with his chicken calls once again.

"Pock-pock, pock-pock, pock-POCK!" he said. "Come on, I'll match everything you have, and even put in an extra five marbles. Hell, I'll make it ten!"

I licked my lips. This was really tempting. But still, I just knew that this wasn't right, because Gus was now moving me from our regular game into a game that I knew nothing about. Still, with the thought of those ten extra marbles, I finally cracked and said, "Okay."

Hearing this, Gus now moved in with the confidence of a panther going after a deer, and he beat me out of almost all of my marbles in one, two, three. Because he wasn't just bigger and stronger than me, but also knew how to work the marbles close to the edge by shooting into three or four marbles at a time. That afternoon, everyone applauded Gus, including the two girls. But then, the girl with the large, kind goat eyes actually came close to me.

"You did really well, too," she said in a kind tone of voice.

My whole world stopped. My God, this was the first girl who'd ever spoken to me in all of my years of going to school. Her name was Nancy and she had long dark brown hair. She offered me a piece of bubble gum, and when I took off the wrapper, she took the funnies from me that were attached to the inside wrapper, and she read them and laughed. I could see she was really smart, because she'd understood the joke—which I never did.

Just then, a breeze picked up, and I glanced up at the huge old eucalyptus tree above us and I could see that he was smiling. I smiled back, thanking the tree. "Always remember," *mi mamagrande* had explained to me many times, "the whole Spirit World is always with us wherever we go, if we just have the *ojo*-eyes to see from our hearts and the *oreja*-ears to hear from our soul."

That afternoon, Gus went home telling everyone how he was still the champ and would always be the champ. And that night, I felt kind of stupid, but still I explained to my brother what had happened. To my surprise, Joseph didn't get mad at me. No, instead he said, "I'm glad that Gus beat you."

"But why are you glad that he beat me?" I asked, feeling all confused that my own brother would say this to me.

"Because you were getting cocky," he said.

"No, I wasn't," I said quickly.

"Look," he said, sitting up in bed, "if you hadn't won all week long, would you have ever agreed to play in that larger circle?"

"Well, no, I guess not."

"Exactly. You agreed to do it because you were beginning to think that you were so good that you didn't need to stick to basics. *Papa* always says that the easiest guys to con are other con artists, because cons are always looking for shortcuts and easy ways in life.

"Now, just let me get out of bed and you show me how large that circle was." I could see that it was hard for my brother to get himself out of bed. "He was smart, you know, to move you from your game to his game. I bet he's had coaching."

"Yeah," I said. "I thought of that, too."

"Good. I'm glad you're thinking. *Papa* always says that knowing how to think and keep his eyes open is what saved him and his mother and sisters time and again during the Revolution."

Going out the back door of our old ranch house, we crossed the field to our new home, which was almost done. My brother had to stop several times to catch his breath. And yet, when we'd go to the Army Navy Academy to watch him play football, we'd see them hand him the ball and he go running real fast, knocking people out of the way.

In the dirt in front of our new two-story home, I drew a circle about the same size that Gus had used to beat me. I couldn't believe it, my brother now quickly showed me how to work a big circle. In a large circle, unlike a smaller circle, you shot into little groups of marbles that were near the edge, not the singles, which you could miss because of the greater distance. And I could see that this was exactly what Gus had done.

Then, I'll never forget, going back to our old home that afternoon, my brother went over to the drawer where he kept his socks all neatly put away—military style, one halfway inside of the other—and he brought out a black sock and turned it upside down on the bed. And here, before my eyes, I saw the prettiest, clear-colored cherry-red marble that I'd ever seen roll out of the sock. It was a normal-size marble, too, but just a little bit larger.

"Back in the *barrio* this was my main shooter," said my brother. "But I didn't use her every day. I used Big Cherry only for special games. She's yours," he added.

"Mine?"

"Yes. Yours. And with this marble, I want you to beat Gus. Big Cherry has been in many battles. You look at her carefully and you'll see she's pretty marked up. Once, one damn—I mean, blessed—*vato* sneaked in a steelie on me in a game of chasies. Don't ever play chasies with her. She's your thoroughbred. You don't use her for roping or bulldogging. Go on, pick her up."

But I couldn't. All I could do was stare at this most beautiful marble that I'd ever seen.

"Take her," he said again.

Finally, I reached down and picked Big Cherry off the bed, holding her between my thumb and index finger so I could hold her up to the light. "But this marble is so beautiful, and I thought you told me that a shooter shouldn't be beautiful," I said.

"There are exceptions to every rule," said my brother. "And besides, feel her weight and surface."

"My God, you're right. She's heavy and she's not too smooth either."

"Exactly. So it's okay if a shooter happens to be beautiful. The main thing is that you didn't pick her for her beauty any more than *papa* chose *mama* because of her great beauty."

"*Mama's* beautiful?" I asked.

"Are you blind?" said my brother. "Don't you see how men and women are always looking at our mother everywhere she goes. Our mother is more beautiful than any movie star you'll ever see, and yet this isn't why *papa* chose her. He chose her, he's told us a thousand times, because when he first saw her, she was in line to go into the dance hall over in *Carlos Malo* with her brother and sister. And when a fight broke out, she didn't get excited and enjoy it like her sister and the other women. No, she and her brother moved away, wanting no part of it, and so, by seeing this, *papa* knew that she was a woman of high intelligence, respect, and responsibility; a person that he could trust with his life. And trust, remember, is the foundation of all love, *papa* says."

"I see," I said. "Then our *mama* is beautiful," I said, big tears coming to my eyes. "And *papa,* he's smart, because he knew how to choose a wife, and knowing how to choose a wife is the most, most important thing that any man can do in all of his life."

"Exactly," said my brother, "because from our wives come—"

"Come our children," I said as I'd heard our dad tell us a thousand times. "And our children inherit their instinct for survival from the women."

My brother smiled. "I see that you've been listening. Good," he added.

I couldn't stop crying. All these years since I'd started school, I'd thought that my mother was ugly and my father was a fool. My brother took me in his arms and held me a long time, and it felt so good.

The following week, I went to school with a power inside of my heart-*corazón* that was BURSTING! And I beat the hell out of everyone at recess, including Gus, with our regular-size circles. And then, after school, when we went out once more to play under the huge old eucalyptus trees, I brought out my brother's Big Cherry and I beat the crap out of Gus with his big circle, too.

And when he said that Big Cherry was the size of a boulder and I couldn't use her unless he could do switchies, I almost said, "Sure, do your switchies! I'll beat you anyway!" But I didn't, because I could now see that this would be cocky talk. I put Big Cherry away and still I was able to beat him with my regular shooter.

Something had happened to me. It was like I just couldn't do anything wrong. But then, I remembered that my brother told me that our dad had learned from Duel up in Montana to always let the other guys win back a little bit of their money at the end of the night of poker-playing, so they'd want to play with you again the next day. I did this. I missed a shot, pretended like I was all upset with myself, and Gus got excited and came in and shot the hell out the last few marbles in the pot.

Then he screamed with joy, and went home that afternoon telling everyone that he was still the real champ, because he'd beat me in the end and he also had a hell of lot more marbles than me.

By the end of that week, it was easy to see who was the new marble champion of our whole third grade, if they just looked past how many marbles we brought to school. But I said nothing. In fact, I could now see that my brother had been very wise to tell me that a real king . . . kept his reign a secret.

I'd listened, practiced, learned, kicked ass, and . . . and . . . I'd also found out that my mother was beautiful and my father was smart and that maybe I wasn't very good in the classroom, but still, I . . . *un Mexicano,* could think, organize a vote, and kick ass at recess!

I couldn't stop crying, I was so happy—I was *El* KING!

CHAPTER nine

Knowing that I was king felt pretty damn—I mean, blessed—good! And now that I didn't feel terrified of going to school, I quit wetting my bed. My days at school began to be magical, and not just at recess, either. In the classroom, I could now see that there was value in some of the things that were being taught to us. It was at this time that we were taught something called multiplication and division. Some kids were having real difficulty catching on, but for me, on the other hand, I was able to catch on how to do my multiplication and division right away.

Different than with the alphabet and reading, numbers made a lot of sense to me. With adding and subtracting I could figure out how many bales of hay we got off a hillside or how many marbles I had. And with multiplication, I could line up my marbles in lines of tens, then just multiply by ten how many lines I had. This was really great. It saved me a whole lot of time, especially now that I had well over four hundred marbles.

Also, different than with the alphabet, in math you only had nine numbers that you had to memorize unless you counted in the zero. And counting in the zero made a lot of sense to me, too, because after you got to "9," then all you had to do was just put down a "1," like you'd done at the beginning and add a "0" to that "1" and you now had "10," which was like a whole new start until you got all the way up to "19." Then all you had to put down was a "2"—which made sense since "2" came after "1"—and now you'd put a "0" behind this "2" and you had yourself "20."

I loved this kind of clear, easy thinking. It made a lot more sense to me than that damn—I mean, blessed—alphabet, where you had to memorize twenty-six letters, then put those letters into bunches with no rhyme or reason to make up different words. Like the word "read," itself, was sometimes "reed" and then other times "red." This pissed me off! Reading and writing were a pain in the ass! Why couldn't we have a lot less letters, and maybe even a "zero" letter to bunch things up with.

Because adding and subtracting made so much sense to me, I quickly learned how to add and subtract two or three columns in my head. None of the other kids could do it. Suddenly I was smart. I began to think that my days as a "slow learner" were behind me and no one at this new school was ever going to find out that I'd been called "stupid" at the other school across town. One day, during reading, our teacher asked who would like to stand up and read aloud, just as our teachers had done in the first and second grades.

My God, the terror that shot through my whole body made me sick. And suddenly, I wasn't king anymore. Hell, I wasn't even a third grader. No, I was a heart-pounding kindergarten-baby-born-in-the-gravy and ready to pee in my pants again.

I'll never forget as long as I live how our teacher looked around the class to see whom to call on first, and when her eyes came towards my section of the classroom, I closed my eyes and said a quick little prayer. "Please, Lord God," I said, "don't let her call on me, because if she does, I'm sure to pee in my pants."

Then I heard her laugh, and so I opened up my eyes, and miracle of miracles, God the Merciful saved me, because instead of needing to pick someone to read, all these hands went flying up, especially the hands of girls, wanting to read.

"Thank God for all girls," I said to God, and I thought that my prayers were answered and all my problems were over forever and ever, and our teacher was only going to ask who wanted to read for the rest of the year. After all, she was right to do this, I figured. We weren't first or second graders anymore. No, we were big third graders, three years away from being "babies-born-in-the-gravy," so reading should be voluntary.

But then one day, she asked, "Well, who hasn't had a chance to read so far?" and she glanced all around the classroom and once again her eyes—oh, my Lord God—came to rest on me.

I got so scared that I thought of bolting from the classroom, jumping on my Schwinn, and pedaling for home as fast as I could. But then, thank God, the bell rang and it was lunchtime.

Quickly, I tried to figure out what to do. I only had so much time and I knew there were quite a few of us who hadn't read yet, because it always seemed to be the same six girls and a couple of boys who always raised their hands, wanting to read.

I went around and very quickly asked the different guys—who I thought hadn't read—if they'd read so far. A couple of these guys immediately said that they hadn't. A few others wanted to know why I wanted to know. I didn't know what to say because, up to that point, everyone in my class assumed that I was one of the smartest and most capable of all us kids, because of my marble shooting and how good I was in math.

"Look," I said, suddenly remembering how my brother Joseph had explained to me that our dad had paid that Acuña kid to teach my brother how to play marbles, "I'll pay you each, well, a nickel if you put up your hand and tell our teacher that you haven't read yet."

"You got a nickel?" asked one kid suspiciously.

"Well, no, not right now, but I'll bring some tomorrow."

"And how do we know that you'll keep your word?"

"Because I promise," I said.

"But everyone knows that Mexicans never keep their—"

This kid never got to finish his words. I was on him like a swarm of flies on fresh shit, knocking him to the ground and hitting him as fast as I could. "Okay! Okay!" he yelled. "I'll do it! I'll do it!"

My heart wouldn't calm down. I wanted to keep hitting him. It was my first real fight. I could now see that all my other fights had just been wrestling matches. This one had been the real thing. I'd really wanted to hurt him just like his words had hurt me. He'd never had a chance. I liked this. Never again would I ever let anyone say anything bad about me or my people. I'd knock the living hell out of them. And if they were bigger than me, then I'd go home and get myself some dynamite and blow them to smithereens!

Suddenly, I remembered Ramón—my hero. And I wished that he'd been here to see me in action. I wondered what had happened to him. He'd been so smart and capable, but our teachers had never seen this. All they'd ever seen was a "stupid, lying, no-good sneaky Mexican."

After lunch, we went back into the classroom, and when the teacher asked who hadn't read yet, my "good friend," whom I'd just beaten up, glanced at me, then put up his hand up real fast, and he read really well. Then another "friend" of mine raised his hand, too. At lunchtime, I'd made deals with four kids, but I'd only had to knock the hell out of one. And the next day, I brought a bunch of nickels in my pocket and paid off the four guys, even the guy whom I'd beaten up. But he didn't want my money.

"Go on, take it," I said. "And no hurt feelings. Just never insult me again."

He nodded, taking the nickel. "I'm sorry," he said. "I didn't know you'd get so mad."

"Well, now you know."

"I didn't even know if you were really a Mexican. I thought all of the Mexicans lived across Oceanside in *Pas-olie* Town."

I never heard the word *"pozole,"* which was a type of Mexican soup, pronounced the way this boy had said it. It sounded vulgar, just like the word "Mexican" had sounded so dirty when that playground rooster-female teacher had yelled it at us all through kindergarten.

"The word is *'pozole,'"* I said to him. "Not *'pas-olie.'* That sounds ugly. And I want you to know that *Mexicanos* do keep their word. In fact, *hombre de palabra,* 'man of his word,' is the highest praise you can give to any man, my father always tells us."

"Okay," he said. "I didn't know."

"Okay," I said, "but now you do know. And yes, I'm Mexican. Both of my parents came from Mexico."

It was over. I could just feel it. Having beat this kid up so easily had opened up a whole new world of possibilities for me. Having worked on the ranch and needing to move pigs and calves and horses around all my life had made me strong and taught me how to maneuver in a way these city kids knew nothing about.

It was almost the end of the year and our teacher asked Gus and me and this one other kid to stay in the classroom during lunchtime, because she needed to speak to us. Instantly, I knew what was going on. We were the three kids who always tried to avoid reading. But I'd seen Gus and this other kid read a couple of times, so I knew that they were a whole lot better than me.

So when lunchtime came, I acted like I'd forgotten what our teacher said, and I got up and started going out with the other kids. Because once I was out the door, I already knew what I was going to do. I was going to jump on my Schwinn bike and pedal home as fast as I could and never come to school again. I'd hide in our big avocado orchards all day, or down in the swamp, then come home in the afternoon and pretend I'd been at school all day.

But I didn't even make it out the door. Our teacher caught me. Her name was Mrs. Morlo, and she wasn't really mean like the other teachers that I'd had. In fact, I'd say that she was the first fair-minded teacher that I'd ever gotten.

"Excuse me," she said, "but I think you forgot that I said I need to speak to you."

There was nothing I could do, she'd caught me cold. Gus and the other kid were grinning. They'd been smart enough not to try to make a break for it. She asked Gus and the other boy to wait for her outside the classroom door, and they went out with the rest of the kids. Now I was alone with Mrs. Morlo.

"I don't think that you've read for me all year, have you?" she said.

I was looking down at the floor, studying the different squares of the tile-like flooring. I'd never realized until now that this was the exact same flooring that they used at Penney's. Remembering Penney's, I began to wonder if I needed any new clothes. Or maybe, now that I was older, I should ask my parents to take me to Sears in Escondido. That's where my father had just bought some great new spurs.

"Look," she said, when I didn't reply, "I saw how quickly you were able to learn your multiplication and division, so I think that you probably just need a little help with your reading. Here, sit down with me," she said in a kind voice, "and let me see you read this page."

I sat down. There was nothing else I could do. I took the book that she gave me, and immediately all these white rivers began running up and down the page between the words. And the words, they would just jump up off the page at me.

I took in a big breath, held, and I tried to focus my vision to just one word at a time, and then move my eyes from the left side of the word to the right side of the word, as we'd been told to do in first and second grades. But it was really hard for me to hold on to the different letters, with all the ups and downs of short letters, long letters, and those other ones with the tails that hung down below the line. I was so scared to make a mistake, and maybe say something in Spanish, that I finally just began crying.

"You can't read, can you?" she asked.

"No, I can't," I said, wiping my tears.

"How long has this been going on?"

"Since we first started reading."

"In first grade?"

"Yes."

"And no one noticed this before?"

"No, ma'am."

She looked at me for a long time, and I could see that her look wasn't mean and she wasn't going to slap me on the head for not paying attention, like my first- and second-grade teachers used to do. I could hardly remember anything else about the second grade, except our teacher, a woman, hitting us Mexican kids so hard and calling us "stupid Mexicans" that my ears still rang. But we hadn't even all been *Mexicanos*, and so we'd sometimes laugh when we'd see her hit the two local

Indian kids who weren't Mexicans, because now, finally, some one else was getting it besides us *vatos*. This was democracy, Ramón told us, everyone getting hit equally.

"Look," she finally said in a kind voice, "I think that if you stay with me a few minutes after school every day, we can have you reading between now and the end of the year. Because, you see," she added, "if we don't have you reading at a third-grade level by the end of the school year, you'll have to repeat the third grade again."

"Repeat?" I said.

"Yes. We'll have to hold you back a year."

I nodded. I understood. I was going to be roped, tied, and branded a "stupid, slow learner" for life if I didn't learn to read well in the next month and a half.

She wrote out a note and told me to take it to my parents so that they'd be informed and know why I'd be getting home a little late every day. Then she told me to go outside and tell Gus to come in. When I went out, Gus gave me a great big grin like I'd never seen him give before.

"So we got busted, eh, partner?" he said. "But it ain't so bad. Hell, I've flunked before."

"You have?"

"Hell, yeah! Back in Texas, I flunked-me one grade real good!"

And saying this, he went in to see the teacher with a swagger like he was king of the world. Suddenly, I wondered what the hell was going on. Ramón had been the smartest and most capable guy in our whole grade in "thinking" and "figuring" out things, and he'd been considered dumb, too. And now Gus, he was also the smartest guy of all of us during recess and he just said that he'd been flunked before. This didn't make any damn—I mean, blessed—sense to me at all! The smartest kids I knew were all considered stupid.

That night, I took the note to my parents and my mother read it and asked me what was going on.

"I don't know," I said, lying. "Tell me, what does the note say, *mama.*"

"Your teacher says that you need help with your reading and that she's willing to stay after school and help you a little each day. She also says that you do very well with your arithmetic, so she thinks that you'll catch up by the end of the year. Do you have trouble reading, *mijito?*" asked my mother.

My eyes filled with tears. I didn't know what to say. I didn't want my own mother to find out that I was stupid. She took me in her arms, and that night she sang to me an extra-long time about God's turtledove *de amor* that says, *"Coo-coo-roocoo-coooo!"* I loved my *mama* so much! Nothing bad could ever happen to me when she held me in her large warm arms.

All that week, Gus and I and this other kid stayed after school and our teacher tried to help us with our reading. Gus was actually pretty good at his reading, and so, by the end of two weeks, he didn't have to stay after school anymore. By the third week, I could see that our teacher was getting short-tempered with the other kid and me. After about thirty minutes of helping us, she'd begin to look at the clock on the wall and get irritated with us.

Then in the fourth week, I don't know what happened, but the other kid quit coming to school altogether, and I was now alone with Mrs. Morlo, and this was when she told me that my parents would have to come to school so she could speak with them in person. It was only two weeks until the end of school.

That day, I went home on my Schwinn bike, crying the whole way. I knew what was coming, and I'd tried to pay attention so hard and learn how to read, but I couldn't do it. All those letters just kept jumping around, making designs that made no sense to me.

At home, I gave my parents the note my teacher had given me to give to them, then we went to the school in my parents' new car. It was a great big, long, navy-blue Cadillac, the first Cadillac in all of Oceanside and Carlsbad after the war. Also, our new home was finished, and now my father and mother were going to downtown San Diego and ordering custom-made furniture from a great big, fancy furniture store that had the biggest elevator I'd ever seen.

I could see that my teacher almost shit in her pants—I mean, dress, when she saw my father and mother drive up in their big car and get out, all dressed up. I don't know what she'd been expecting, but here was my father dressed in tailor-made silver-gray Western slacks, a beautiful handmade belt and buckle from Mexico, and his big cowboy hat with the fancy horsehair band. And my mother, she was dressed in her Sunday best, looking like a movie star. Myself, I was still in my school clothes that were the same ones that I used for working around the ranch.

Nervously, Mrs. Morlo fixed her hair, then kept asking my parents to please sit down until she realized that our third-grade furniture was too small for adults. Still my father maneuvered his large, thick body into one of the little desk chairs. It was really funny, seeing *mi papa* in one of our desk chairs. My mother didn't even try. Calming down, my teacher finally told my parents the same thing that she'd told me, that I was smart at arithmetic, so . . . she'd thought that I could catch up on my reading, but . . . I hadn't been able to do it. Then she dropped the bomb.

"I'll be recommending that he repeat the third grade," she said.

I watched my parents look at each other. My mother was the first to speak.

"Does this mean that he'll be held back a year?" asked my mother.

"Yes," said my teacher. I could see that she felt pretty terrible, herself.

"I see," said my mother, and she now also tried to put her body into one of our small desk chairs as my father had. I could see that she was thinking and trying to understand the situation. "Are there any other children who are being held back?" asked my mother.

Hearing this, my father winked at me. I guess that he thought that this was a good question and he was proud of my mother for having asked it.

"No," said my teacher. "The other boy, whom I was also going to recommend to be held back, he and his family have moved from the area."

I didn't want to, but I couldn't help myself, and I began to cry. Then I was going to be the only "stupid, slow learner" to be held back in all of my class. I'd so much hoped to spare my parents the terrible shame of finding out that their son was stupid. My brother Joseph and sister Tencha had never flunked. I wished I could die, I felt so bad.

Seeing the tears running down my face, my father put his huge thick hand on my shoulder, stroking me gently. "Tell me," he said to my teacher, "just how much you want?"

"How much what?" asked my teacher.

"Money," said my dad, reaching for his pocket. "We can settle this thing right now."

"Salvador!" said my mother. "Please, none of that!"

"*¿Y porque no?*" he said, getting out of the little desk and bringing out the huge roll of money that he always kept in the right front pocket of his slacks or Levi's. Only fools carried their money in wallets in their back pockets, our father always told us. "Money talks, hocus-pocus bull-shit just stinks!"

"Excuse me," said my teacher, "but this isn't about money. This is about your son's future. If he doesn't get his foundation of reading down, he'll have trouble all of his life."

"Good," said my father. "All life is trouble, so why the hell worry, eh?"

"Excuse me, but you don't seem to understand."

"I DON'T UNDERSTAND!" roared my father, putting his money back in his pocket. "Hell, I've forgotten more than you or most people will EVER UNDERSTAND!"

"Salvador," said my mother as quietly as she could, "why don't you and *Mundo* go outside and let me talk to this woman alone."

"Damn good idea!" said my father. "Let's go outside, *mijo*, and get us some fresh air. It stinks *de puro pedo* here!"

"Listen to me good," said my father the moment we were out the door. He was hot, I could tell. "Everybody has their own game, under-stand? Lawyers got theirs. Doctors got theirs. Business people got theirs. Every bum on the street has his, too. Got it? And every game has two sets of rules, the one set that they tell people that they play by, but— listen closely—behind their closed doors, these same people always got another set of rules that they really play their game with. The Church, she does this beautifully, having people pray to *Cristo*, oh, so sweetly.

Then they get all those young nuns and priests to work for free for them all their lives, and yet from behind those closed doors, that good-hearted, all-loving Church steals the best lands of Mexico, and the whole world, if she could!

"Education, *mijo,* is another racket. Another con game! Don't let nobody fool you! School wants to get people thinking all the same way like trained mice. Don't you ever fall for nobody's racket, *mijito.* Think, here in your head, feel, here in your heart, and trust your *tanates,* here between your legs *a lo chingón!* This is life in all her power and glory! Got it?" he said, gently putting his huge thick hand on my shoulder.

"I got it, *papa,*" I said, wiping the tears out of my eyes. And I really did get it. I loved my father *con todo mi corazón.* He made so much sense, just like Ramón, and even Gus. All these guys made sense and they took no shit from nobody!

"Good, because knowing how to think and how to work is how you get ahead in life; not just education. Talone, that *Italiano* over from Temecula, who buys our cattle from us, he says that he had to lose his leg before he finally learned how to think. Before that, he was always strong like a bull, trying to move the world with his muscles. But then, he lost his right leg in a tractor accident, couldn't be hired anymore, and he began to think. He bought a calf here, a goat there, butchered them and sold the meat. Soon he was buying two, three head of cattle at a time and supplying the best restaurants in San Diego with steaks, and selling the rest of the meat out of his butcher shop at a good, fair price. He's a multimillionaire now. He has no school education, but he has life education that taught him to open his eyes, see, and think. You're doing good, *mijito.* You're a worker, and I've seen you think and figure things out. This woman teacher says that you'll have trouble in life. Well, I say good! Let trouble come, and you just kick trouble's ass *a lo chingón, eh?*"

Suddenly my whole heart felt full of power. My dad had turned the whole thing around. I didn't feel terrible or stupid anymore. I was able to stop crying. Some kids now came by on their bikes and they saw my father with his big Western hat, fancy shirt and belt, and cowboy boots.

"Do you really have a horse, mister?" asked one kid, dropping his right foot off the peddle of his bike to the ground.

"Hell, yes," said my dad. "A whole string of them!"

"A string of horses?"

"Yeah, you know, like when you string them up to a wagon."

"You mean you got a stagecoach like in the Wild West?"

"Yeah," said my dad, "a feed wagon."

"Could we come over and ride your horses?" asked the other kid.

"Yeah, sure," said my dad, glancing at me, "if it's all right with my son."

I was so glad that my dad had thought of including me, because these two guys weren't really my friends, so I didn't want them coming to our ranch. But before I could say anything, my mother came out of the classroom. She looked pensive, and all the way back to our *rancho grande,* we drove in silence. At home my mother explained to me that I was going to have to repeat the third grade. And how my classmates found out about this, I'll never know, but for the last two weeks of school, they made fun of me, calling me "a retard, a flunker," and once more school became a living nightmare.

But this time, I didn't start wetting my bed again. No, my father's words had given me strength and confidence. And the two guys who'd asked if they could come horseback riding, I never invited them to our *rancho grande,* but I did invite Gus. And Gus, I found out, really had flunked a grade in Texas, and so this was why he was bigger and stronger than the rest of us kids. He was a whole year older than the rest of us.

It turned out that Gus also knew a lot about horses. He was able to saddle his own horse and he was a pretty damn—I mean, blessed— good rider, too. We ended up playing cowboys and Indians on horse-back, running through different orchards. I was the Indian. He was the cowboy. And one day I tried jumping off my horse to a tree branch at a full gallop, so I could swing up into the tree as I'd seen done in a cow-boy movie in which the hero had swung up in the tree and then jumped out of the tree on the bad guy who was coming behind him. But I never made it. Hitting the branch, I got knocked backwards off my horse so hard that when I hit the ground, I saw stars.

The next thing I knew, Gus was on the ground beside me, slapping me in the face and asking me if I was dead. And when I finally came to, he started laughing and laughing, saying that he'd never seen someone do such a crazy-ass stunt in all his life!

"But I saw the hero do it in a movie," I said.

"Yeah, but movies are fake, didn't you know that? In the movies, I'm sure they don't really run their horses under those trees at a full gallop. They probably just trot their horses under, then the movie maker speeds up the film, you fool!"

That was the day I realized that movies weren't really true. And yeah, sure, I'd always known that movies were pretend, but I'd, also, kind of thought that they were real in how they were made and that the movie stars were kind of, well, really like those people that they pretended to be. I'd almost had my head knocked off when I'd tried to do what I'd seen them do in the movies. I was never going to trust these damn— I mean, blessed—movies again!

One day, Gus and I were walking home—he had no bike, so I'd stopped riding mine so he wouldn't feel bad—and it started sprinkling and all these snails came out on the road. I don't know why, but Gus and I began leaping on them with our boots and smashing them into the ground. Dozens of them! Hundreds of them! Squish! Squish! Squish! Popping their little shells!

Just this side of California Street, we spotted a great big lizard and we caught him. It had stopped sprinkling, so we found some newspaper, set fire to the paper in the middle of the dirt road, and tossed the lizard into the flames.

We watched the lizard trying his best to get out of the burning fire, but every time he'd almost make it out, we'd kick him back into the flames with our boots. Once, I'll never forget, the lizard opened his mouth and I'd never heard a lizard scream before. It was awful. It almost sounded like a little human baby crying. That night it rained and stormed and I was afraid to go to sleep, because every time I'd close my eyes, a great big lizard would come out of the storm to my window and open his mouth, screaming out in agony!

I got up and ran out of my bedroom, yelling for my mother. Finding her, I told her that there was a huge lizard at my window. "Bigger than a horse!" I shouted.

"*Mijito,*" said my mother, "there are no such lizards. You're just having a bad dream."

My parents were having dinner with our good German friends, the Huelsters, and I was supposed to be asleep by now.

"No, *mama,* I'm not dreaming," I said. "My eyes were wide open and I could see the lizard! Come, I'll show him to you. When he opens his mouth, it's as big as lifting the hood of our new car."

Hans and Helen Huelster laughed. They were my little sister Linda's godparents and they lived out at Bonsall where they had a huge chicken ranch, and they'd come into town to see us.

"Excuse me," said my mother to Hans and Helen, "I'll be right back." And she got up and took my hand and walked me down the long dark hallway of our old ranch house to my bedroom. Ever since my brother Joseph wasn't feeling well, I'd been sleeping alone in a room in the back. None of us were allowed to disturb our brother Joseph, who needed all the rest he could get.

But when my mother and I went into my bedroom, the huge lizard was gone. He wasn't outside of my window anymore. Now all I could see outside of my window was the storm and lightning and the hanging branches of our old huge, persimmon tree that was dancing in the wind to beat hell.

"But he was right there, *mama!* I swear it! Right there, in front of those branches!"

"It was just your imagination, *mijito,*" said my mother.

"No, *mama!*" I yelled. "I really, really saw him! And he was huge, and he opened his mouth so big that I thought he was going to come in and eat me and the whole house!"

"Get under the covers, *mijito*," she said, "and I'll sing you to sleep."

Hearing my mother say this, I instantly felt all better and quickly got under my covers. *Mi mama* began to sing to me *"Coo-coo-roocoo-cooo,"* as she gently soothed my forehead and tummy. Everything began to feel warm and good and safe and wonderful. Soon I must've fallen asleep, because no sooner had my mother left, and here was that huge monster at my window once again!

But this time he was a frog. A great big, huge frog! And when he opened his mouth, croaking real loud, his long tongue came out, throwing up dead snails and burnt lizards all over me on my bed.

I SCREAMED, leaped out of my bed, and went racing out of my room, down the hallway, and to the dining room table where my parents were now having after-dinner drinks and coffee with Hans and Helen Huelster.

"MAMA!" I screamed. "HE'S A FROG NOW! And he threw up dead snails and lizards all over me!"

My mother glanced at Hans and Helen. She looked kind of embarrassed. *"Mijito,"* she said, "there are no giant frogs or lizards. It's all just your imagination."

"No, Lupe," said my father, "the boy is right. I've seen these monster frogs and lizards, too. Even ants once, as big as trucks. Come here, *mijito*," he said. "Tell me," he added, holding me at arm's length by my shoulders as he looked into my eyes, "man-to-man, what terrible thing did you do to some snails and frogs today?"

I almost shit in my pajama bottoms. How did my dad know this? And if he did, did he really expect me to tell him in front of everyone what it was that Gus and I had done. We'd been awful! And that poor lizard, I could still see his mouth wide open as he screamed for help.

"Come on, *mijito*, I've told you a thousand times, that to the cop or priest we might need to lie, but to ourselves or *la familia*, we never lie. What did you do? Did the snails come out by the hundreds this afternoon in the rain and you went around killing them?"

I had to hold myself with both hands between the legs real tight so I wouldn't pee. My God, how did my dad know this? "Yes, *papa,*" I finally said. "We . . . we did."

Hans laughed, saying, "Good riddance! Damn snails! They get into everything! But on our place the chickens clean them out!"

"And that's the way it should be, Hans," said my father to the big German. "Nothing wasted. Just a normal part of life. Chickens eating snails, but here . . . we have a whole different situation. These boys were killing just for fun."

Hans laughed. I could see that he was feeling nervous. "Sal, he's just a boy," said Hans, "and boys will be boys, you know that."

"No, Hans, boys do not have to just be boys. That's why my mother raised me up like a woman for the first seven years of my life, so that I, a male, would then have respect for all life just like women, who know that all life is sacred. Tell me, *mijito,*" said my father, turning back to me, "and the frog? What did you boys do to a frog?"

"Nothing to a frog, *papa,*" I said, beginning to cry. "It was a lizard that we caught and we . . . we burnt him."

"Alive?" asked my dad.

"Yes," I had to say.

My father nodded, and then nodded again. And when he reached out for me, I ducked, thinking that he was going to hit me. But he didn't. Instead, he took me into his arms, and held me close. I could feel his whole chest expand and contract against me as he breathed.

"Well then, you got good reason to be seeing all these monster lizards and frogs and snails, don't you?" he said. I nodded. "So now you got to go to your room, *mijito,* and kneel down and pray and ask *Papito Dios* to forgive you. Because, remember, as I've taught you and your brother here on the ranch, we never take any life without showing respect. And above all else, we never torture or have fun in the act of killing. Got it? We know what we do, and we do it quick and gently and always for a good reason. To kill even a snail, just for fun, is not being respectful of—"

"Of God, Who is in all things," I said as I'd been told all of my life for as long as I could remember.

"Exactly," said my father. "So now go and pray and ask *Papito* to forgive you for having offended Him."

I nodded and I glanced around. Everyone was looking at me. "But I'm scared," I said. "Will you go with me to pray, *papa?*"

"No, this you must do alone, *mijito,*" he said.

"But he says that he's scared," said my mother. "And he's just a little boy, Salvador."

"How old are you?" asked my dad.

"Eight," I said.

"You see, Lupe, he's not a little boy. He's eight years old. One year past being seven years old, and at eight is when we stop raising a boy as a girl and we step him through that needle's eye . . . into manhood. He was big enough to kill and torture without mercy, Lupe, and so now he needs to earn his *tanates,* because *tanates,* without respect, are very, very dangerous.

"And also, you tell me, Lupe, what's wrong with him being scared? Eh, the deer lives her whole life in fear of the lion, but still she has her babies and a good life. Fear is good, Lupe. It helped you and me be strong in the Revolution, and now in peacetime, fear is the only thing that helps keep us, people, honest, especially men. Now, go on alone, *mijito,* and you ask *Papito Dios* to help you make peace with these monster snails and frogs. Then they'll leave you alone."

Still, I didn't want to go alone. No, I wanted to leap into my mother's arms and have her hold me forever and ever. But I could also see that my dad was right, and I needed to walk down that long, dark hallway to my room by myself. I hugged and kissed my father that night as I'd never hugged and kissed him before, then I hugged my mother real tight, kissing her, too. Then I said, *"Gute nacht,"* in German to Hans and Helen, hugged them also. Then I turned and started down the long hallway, feeling more scared that I'd ever felt in all my life.

But I could really see that what my father had said to me made a whole lot of sense, and in doing this alone, by myself, I was on my way to becoming *un hombre a las todas,* a man for all seasons, which meant, literally—as my dad had explained to my brother and me many times—a man whose balls had dropped so he could then reproduce in any kind of weather, hot or cold, all year long. Because males, whose balls hadn't dropped, could then only reproduce during a few days in the spring of each year. And fear was the main reason why most men's balls remained hidden up in their body.

I could now see very clearly that Ramón, even though he'd just been a little kid, had already been *un hombre de las todas* with dropped balls in kindergarten. And my father, he'd been forced to become this type of man at ten years old in the Mexican Revolution. Suddenly, I realized that Jesus Christ must've really had big dropped balls or He would've screamed out in fear when they'd drove those nails into Him. Yeah, sure, this made sense. He would've never been able to say, "Forgive them for they do not know what they do," if He hadn't had dropped *tanates,* and big ones, too.

Thinking this, I felt a whole lot better as I walked down the long, dark hallway. Then I heard the floorboards of our old house screeching beside me as I went barefoot down the long hallway. I smiled. Jesus was here with me. I could feel Him. I wasn't walking alone.

Getting to my room, I immediately saw that there was no huge lizard or frog at my window. So I went inside, leaving the door open, and knelt down before my bed and I began to pray . . . as I'd never prayed before in all my life.

"Dear *Papito Dios,*" I said, "please, I ask You with all my heart and soul to forgive me for having killed all those snails and that one lizard. I'll never do it again. I swear, so please, please, please, forgive me. In fact, tomorrow, if it rains, I'll pick up snails all afternoon on Stewart Street and help them get across the road so that the cars don't run over them. Please, dear Lord God, I just didn't know what I was doing."

After saying this, that I "just didn't know what I was doing," a bad feeling raced up and down my spine. Then this meant that I was exactly like one of those guys who'd drove the nails into Jesus' hands.

I started crying. "PLEASE, LORD GOD!" I yelled. "FORGIVE ME! FORGIVE GUS and ME! Because I now see that both of us really didn't know what we were doing, or we would've never, never done it! Please forgive us! And I'll tell Gus, too! No more stomping snails with our cowboy boots, EVER!"

It was at this moment, as I said all this, that I heard a huge croak and I turned and saw that huge frog outside of my window again. But this time he wasn't scary anymore. No, now his eyes were smiling and he looked so kind.

"I forgive you," said the frog.

"You forgive me?" I said. "But I was talking to God."

"Yes, I know," said Frog.

At that very instant, I finally understood everything that *mi mamagrande* had been telling me ever since I was little and we'd lived in the *barrio de Carlos Malo,* meaning Carlsbad, and she'd taken me hand in hand all through her Garden. Because I could now see so clearly that this Frog was the Animal Spirit that had been given to me at birth.

And the Animal Spirit that was assigned to a person when they came into this world was, indeed, the Window through which they could see God. And those Snails that Gus and I had killed were God, too. This, my Animal Spirit, Frog, was now telling me.

All of Life was Holy, Holy, HOLY, and SACRED! This was why my brother Joseph had explained to me that our *mamagrande,* Doña Margarita, had never allowed him and my sister Tencha to "damn" anything. To "bless" ourselves and everything was the only way to live, I now saw this so "blessed" clearly!

"Goodnight, Frog," I said.

"Goodnight, my Child," said Frog.

That night in my dreams, I had wonderful travels. I drew a Star in my sleep, colored it in, then jumped in the Star and I was off to Heaven! I'd prayed, found my Animal Spirit, and so I was able to sleep beautifully all night long up in *Papito Dios'* Holy Arms.

In the morning, I awoke feeling wonderful, meaning full-of-wonder.

I was eight years old, one year past seven, and I'd done as my father had told me to do. I'd gone down that long, dark hallway alone, shitting in my pants with fear. But then I'd felt Jesus at my side and I'd prayed, asking for forgiveness, and I'd passed through the needle's eye into manhood.

CHAPTER ten

Our huge house was DONE! FINISHED! COMPLETED! And this afternoon *mi papa* and I were way out by the front gates of our *rancho grande*. He was letting me drive our big new tractor while he hooked up the chain to a tree stump that we'd dynamited so we could drag the stump away. Boy-oh-boy, I now knew what I wanted to be when I grew up! I wanted to be a tractor-driving, dynamite-setting cowboy, then I could just about outdo even Superman!

"Put it in reverse!" yelled my dad. "And come back real slow! That's it! Real slow! I don't want you running my ass over!"

I did what he said real carefully, because I'd seen one of our workmen go to the hospital last week with a crushed leg, because our main tractor driver hadn't been paying close attention.

We were dragging stumps out of the way and getting the front grounds of our ranch ready for our big housewarming celebration. It was almost sundown when a man drove up in a beautiful, long baby-blue convertible. He wore a white panama hat and looked like he'd just stepped out of a magazine, he was so clean. Even his shoes were spotless. He didn't have cow shit anywhere on him, and it wasn't even Sunday.

"Excuse me," he said, getting out of his car and walking up to my father and me, who were covered with dirt and mud and probably some cow manure, too. "But isn't this the place where they're going to have a big celebration this weekend?"

"*Y que cabrones!?!*" said my dad, wrapping the chain around the next stump that we needed to move. We were now working next to the little three-bedroom guesthouse that my parents had built near the front gates of our *rancho grande* so that we could stop people from coming all the way down our long driveway to our main house. My dad and I had been working outside for hours. My mother and brother and two sisters had gone down to San Diego to check on the last of our furniture that was being shipped in from Europe.

"Do you speak English?" asked the man. He had on a short-sleeve shirt with colorful birds and flowers on it, and cream-white slacks and those damn—I mean, blessed—clean two-tone shoes.

"English?" said my dad. "Yeah, sometimes! But then, some other times, no damn way!"

"Oh, I see," said the man, looking confused. "Well, I was wondering if this was where the mansion was built, but this house," he added, referring to the guesthouse, "doesn't look that big."

"Hell, no, this is the guesthouse!" shouted my father, having wrapped and hooked the chain and signaled for me to now put it in low gear and go forward. "The castle IS THAT WAY, past all those eucalyptus and orchards!"

"A castle? Really?"

"Twenty or thirty rooms, and each room big enough to hold a horse and buggy."

"Yes. I'd heard that, too," said the man with a big smile. "Do you work for him?" yelled the man, wanting to be heard above the roar of the tractor as I now went forward.

"WHO?"

"THE OWNER! I hear that he's NOT from THESE PARTS!"

"HELL, NO!" yelled my father. "He's from Europe, I hear. And Mexico, too! He has castles, I hear, ALL OVER THE WORLD!"

"Really?"

"Well, I don't know," said my dad, signaling for me to cut the motor. I guessed that I'd pulled the stump far enough out of the way. "I never seen them, but that's what I've heard!"

"Then you do work for him?"

"Yeah, my son and me, but we've only seen him just a few times. You see, he comes, you know, at night, then he's gone. But we got to keep everything ready for him day or night, in case he and his wife show up."

"I hear tell that his wife is beautiful!" said the man.

Hearing this, my dad turned and looked at the man, sizing him up. Myself, I could hardly keep from laughing. My dad turned and gave me a hard look. I stopped my smiles. "Get this damn tractor to the barn!" he barked at me. "That's it for the day!"

"Look, I'm sorry I'm disturbing you," said the man. "I don't want to get you and your son fired," he added.

"You won't," said my dad. "The owner, he's not here right now. Come on, I'll show you the place. And yes, I saw his wife once. Not from real close, but what I saw, I'd say that yeah, she's more beaut-eee-full than any movie star you'll ever see!" said my dad with pride.

"Really?" said the man.

"Yes. Now, come on, you give me a ride in your car, and my son will take the tractor in for the day."

"You trust a boy with that huge tractor?"

"Hell, I trust that boy with my life!" yelled my father. "He can out-ride, out-think most grown men! And next year, he'll have his own gun, so that no son-of-a-bitch man can walk on his shadow without permission!"

"A gun? But he's a child!"

"It's never too early for a boy or girl to learn about the ins and outs and responsibilities of life!"

"But with a gun?"

"Shit, yes, with a gun?" said dad, opening the passenger door to his car.

Suddenly, I could see that the man wasn't too worried about guns anymore. Now he was all upset about my father getting into his nice clean convertible. My dad saw it, too, but he just winked at me and got in the man's car, then dusted all the dirt off himself. The man almost shit a brick. I had to work hard not to laugh.

"Straight ahead, friend," said my dad, acting like he hadn't noticed the man's reaction.

SAL AND LUPE IN LAS VEGAS, NEVADA, 1953.

1954

VICTOR, FOURTEEN YEARS OLD,
STANDING IN FRONT OF A PORTRAIT
OF HIS BROTHER JOSEPH, 1954.

(PHOTO: LINDA VILLASEÑOR)

VICTOR, SIX YEARS OLD, ON
MIDNIGHT DUKE, 1946.

VICTOR, THREE YEARS OLD,
SITTING ON A BAR STOOL AT
SAL'S POOL HALL, 1943.

VICTOR, SEVEN YEARS OLD,
WITH SHEP, 1947.

FROM RIGHT TO LEFT:
CONNIE ALARCÓN, A FAMILY FRIEND;
LUPE; SALVADOR; HORTENSIA;
AND IN FRONT OF THEM, JOSEPH.
ON THE FAR LEFT, VICTOR,
TWO YEARS OLD, 1942.

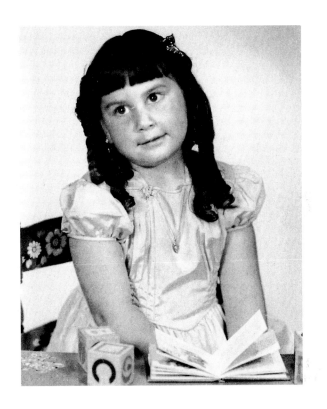

LINDA, 1949.

SALVADOR, HORTENSIA, AND LUPE
ON A TRIP TO MEXICO, 1949.

"Ford. John Ford," said the man.

"*Juan. Juan Puro Pedo,*" said my dad, taking the man's clean hand in his dirty hand.

They drove through our big white stucco gates, passed the Spanish-style guesthouse, and down the long line of towering eucalyptus trees, and passed the orchards of avocados, lemons, and oranges. I put the tractor in third so I could go a little bit faster, and followed them down the long driveway, then took the right fork up towards the tractor barn and horse corrals. They took the left fork and drove up the newly black-topped driveway to our new two-story home.

For over a month, we'd been planning this great big housewarming celebration. We'd fattened a steer, two goats, three pigs, and there were a dozen women working on homemade *tortillas, enchiladas, frijoles, arroz, salsa,* potato salad, corn on the cob, string beans, and fresh apple cobbler *a la capirotada,* with fresh honey. All of Carlsbad and Oceanside were invited. The celebration was to last three days. *Mariachis* were being brought in from San Diego and *Tijuana. Tequila,* beer, and wine were going to be served by the barrel. More than five hundred people were expected. Cars and trucks and horse-drawn buggies were going to be parked in the orchards.

For nearly a month, we'd all been working around the clock, and then last week my mother had told my father that she wanted some white gates to be built at the front of our *rancho grande* to give the celebration an elegant entrance. And my father, he hadn't gotten angry with her request. No, he'd looked at our mother in silence, then he'd turned to us kids.

"Just look at this woman that I married," he'd said to us, "she was born in a little *ranchita* in the mountains of *Chihuahua, Mexico,* and she grew up so poor that they had nothing but a shack anchored to a rock, and here she feels so natural giving orders to everyone like she'd been a queen all of her life."

"Don't ridicule me, Salvador!" snapped our mother.

"I'm not, *querida,*" said our dad. "I'm doing the opposite! Because you just never fail to amaze me! I think I know you, then you become even greater than I DREAMED! I give you honor, *querida!* You are *LA REINA DE MI CORAZÓN,* the queen of my heart! So much blood and hunger we saw, and still you rise up with the vision of an angel! Hell, yeah, we'll get those gates built, *y pronto!"*

I'll never forget the old bricklayer that my dad brought in to do the job. His arms and hands were huge and hairy and he wore his Levi's so low that you could see the crack of his ass. He always carried a pint bottle of whiskey in his back pocket and farted louder than a racehorse when he worked. He was the best and fastest bricklayer in all of Southern California. He was the one who'd done all the brick work of the front and back patios of our new home, and now he went to work on those front gates like a stud horse going after a mare in heat.

"Of course, I'll get these damn big gates done in time for your celebration," he'd said to my dad. "You're my kind of man, Sal! You love your whiskey and love your wife, but not necessarily in that order!"

Also, late one afternoon a stone artist from France came to our house and showed us the pictures of the famous Cliff Gardens Restaurant that he'd built in Los Angeles.

"How long will it take you to build us two water fountains?" asked my father.

"Oh, maybe two or three months," said the stone artist, who told us to call him Frenchie.

"I'll give you two weeks!" said our dad.

"Impossible," said Frenchie.

"Why?"

"Well, just to get the material alone will take weeks."

"Why?"

"Well, because up in Los Angeles we had to wait for this, and then that, and it took weeks to get our supplies because of the war just ending."

"No problema," said my dad. "I have connections that no money can buy! You tell me what you need, and I'll have it here tomorrow!"

The stone artist couldn't stop grinning. "I like your style," he said to our dad.

And so Frenchie moved in with us and worked day and night, eating Mexican food three times a day and drinking whiskey at night, and miracle of miracles, he completed one fountain in the front patio and another in the back patio in three weeks.

But then there was another *problema*. Our mother wanted an elegant housewarming with fine dishes and crystal. But our father wanted a real kick-ass *pachanga a lo ranchero* with *tortillas*, and everyone eating with their hands. After about a week of debating this issue, it was decided that yes, champagne would be served in real crystal glasses, but also when the steer was taken out of the ground pit, the head and brains of the animal would be served with *tortillas* on a platter at the main table in *puro Tapatio de Jalisco rancho grande* tradition!

Parking the tractor in the tractor barn, I quickly ran between the fruit trees to our new home, and there was my father in the front patio, showing John Ford the layout of our new place. We still hadn't moved in all the way yet. We were still waiting for the rest of our furniture to arrive.

"It's even more spectacular than I'd been told," Ford was telling my dad as I came running up. "Is this fountain made of real stone? It's beautiful!"

"Absolutely!" said my father. "Everything is genuine real! In fact, there are two natural springs, one here in front and another one in the back. That's why my boss had his house built here, between the two springs."

"This is a natural spring?"

"You got it," said my dad with a grand smile.

I almost laughed. These weren't natural springs. Hell, the water was piped in with copper pipes and the rocks were fake. But Frenchie, that great artist from France, had made the stones look so real, even though he'd made them out of cement, that it was hard to tell, especially with all the ferns and flowers planted around the little waterfall.

"You want to go inside?" asked my father.

"Oh, no. I've seen enough. I don't want to get you and your boy in trouble."

"I told you, the boss is gone right now. He wouldn't be back till late tonight. Come on in and I'll show you around," said my dad, opening the front door.

The house was still mostly empty and so our footsteps echoed through out the large rooms.

"Beautiful floors," said the man.

"All hardwood, clear-oak," said my dad. "Damn hard to get right now, because it's so soon after the war."

"Yes, I know. Some of my friends up in Los Angeles have been wanting to remodel their houses, but can't get the material."

"Yeah, that's why it took my boss two years to build this house. But he's got big connections in high places, so he pulled strings to get what he needed."

"I understand that he's called the Al Capone of the West," said the man.

My father glanced around acting like he was suddenly real nervous. "Who told you that?" whispered my dad.

"Well, I thought it was common knowledge that he'd been a bootlegger."

"*Ssssh,*" said my father. "I didn't know that. But then, that explains why he always comes with a car full of armed bodyguards."

"He does?"

"Yes," said my father, still acting like he was nervous as hell. "I work for him ten years, but I never ask him no questions. He treats me and my kid good, and that's all I need to know."

"Maybe I better leave," said the man.

"No, come in and see the bar. I'll fix you a drink."

"I'm leaving," said the man.

"Look, I told you, he's not here right now. Come on down the hallway and I'll fix us both a drink."

I watched my father take the man by the arm and walk him past the living room with the new carpet and spotless grand piano, past the dining room, by the kitchen, and to the playroom with the built-in bar. Outside of the playroom was the huge back patio for dancing. Next to it was the second fountain that was surrounded with even greater stonework and a profusion of palm trees and ferns.

"My God, this is a paradise!" said the man.

"Yeah, pretty fancy," said my dad. "What do you want—scotch, *tequila*, whiskey?"

"Do you know if he has any of his old bootleg whiskey left? I hear tell he used to make the best liquor in the west!"

"That's what I've heard, too," said my dad. "Look, I think I know where he just might keep his last bottle of that stuff, but . . . you got to promise to never tell a soul, or I could not just get fired, but, well, you know, if it ever got back to him, my son and I could disappear."

The man's eyes twisted. "Honestly! My God, I won't tell a soul," he said, licking his lips.

"Okay, let me see. Yeah, here's the last bottle left," said my dad.

I had to bite my lip to not laugh as I watched my dad bring out his special old bottle and serve them each an inch of whiskey in a large glass. The man watched my father pour the golden liquid like it was pure gold. And I knew that this really wasn't even my dad's real bootleg whiskey. That, he kept in the cellar under lock and key. What he was serving this man was Sunny Brook, or Kentucky Cream or Canadian Club, which he said were almost as good as his own, and he kept one of these in an old bottle for an occasion just like this.

Having poured the two whiskeys, my dad handed the man one glass and took the other. The man sniffed his whiskey, held it up to see its color, nodded his approval, then he sipped it like it was the greatest liquor he'd ever tasted.

"Ah, that's good!" he said. "The best! But I better get out of here. I don't want to get you and your son in trouble."

"Come this weekend to the celebration," said my dad.

"Really? You think I could do that?"

"Sure. Why not? He's inviting the whole town. It will be a feast for kings! A whole barbecued steer, Mexican-style, and music from across the border."

"Okay. I'll come. And here," he said, reaching into his pocket, "for you and your son." He handed my father a five-dollar bill.

"Thanks," said my father, pocketing the money. "You don't know how much this means to me. *Gracias.*"

"Doesn't he pay you well?"

"Well, you know how rich people are. They didn't get rich by being generous."

"Well, here let me give you another—"

"No, no, this is enough. Thank you very *mucho*," said my father, continuing to act like he didn't know how to speak English very well. "I'll walk you out to your car."

I watched my father walk the man out to his car and see him off. Then my dad came back into the house laughing so hard that I thought he'd choke. *"Lo chingé!"* yelled my father. "And got paid for it, too! Oh, this is BEAUTIFUL! And here, you get half of the five dollars, *mijo*, for not laughing. That was fun, eh?"

"But you lied about everything, *papa?*"

"Yes, and he loved it!"

My dad fixed himself another drink and laughed all the more. And when my mother came home with my brother and two sisters—they'd returned from San Diego—he told them the story and my brother and sisters laughed. My mother didn't.

"And what will you do if he comes, Salvador. You'll look like a fool," said our mother.

My father just winked. "No worry, *querida*, you just wait and see. Hell, he's a big movie producer from Hollywood and he has a beach house down in Del Mar. Hell, he'll probably end up wanting to put me in the movies!"

Early the next morning, we had to slaughter the steer. We'd already killed the two goats and the three pigs and taken the meat to

Talone's in Vista to keep the meat in his big walk-in cooler. Now all we had to do was the steer.

For two months, we'd kept the steer penned up and we'd been grain-feeding him. Now he was over a thousand pounds. Carefully, we got one rope over his horns and another rope on his left hind hoof. This way we could walk him from the corrals down the dirt road to our tractor barn where we had the chain and pulley to hoist him up after we killed him. He wasn't a wild-range animal anymore. We'd befriended him over the last two months while we'd been grain-feeding him in the small fattening corral. He trusted us and so we were able to walk him quietly from the corrals to the tractor barn.

A couple of years back, George Lopez, a big, handsome family friend, had come roaring up on his Harley-Davidson the day we were going to kill a steer and the animal had spooked and taken off racing through the orchard. It had taken all the horsemanship that my father and my brother knew to get their ropes around two tree trunks and bring the bucking steer to a halt.

Now we had a truck parked at the fork in our road that came up towards the corrals so that nobody could come in on us. We were doing pretty well and the big, young, juicy steer was almost inside the tractor barn where we had a little grain and alfalfa waiting for him, when he abruptly stopped and began sniffing the ground. I glanced at my father. I guessed that the animal could smell where we'd killed and butchered the three pigs and two goats. But because of all the gas and oil of our farm equipment, we'd thought that he wouldn't smell the other animals' blood and guts. We were wrong. He was suddenly very wary.

"Hold on to your ropes real good," said my father to my brother and our foreman. "It looks like he's gonna spook."

Now the huge steer was snorting and the hair had come up on his back.

"So what are we to do?" asked my brother.

"I don't know," said our father. "But we got to keep calm so he don't smell any fear coming off of us."

All of our lives our father had explained to us that there was a proper way to kill. That the animal had to be kept cool and calm or the meat would have a funny little smell that wasn't good for our digestion. "That's why the meats from the big markets sometimes don't feel very good to us in the stomach," our dad explained to us. "Because most slaughterhouses don't give a damn how they kill an animal. This is the reason Talone's steaks are the sweetest in San Diego County. To kill with patience and understanding, even in war, is what my grandfather Don Pio, the greatest man who's ever lived, explained to us was necessary for any good leader."

We all knew the story of our great grandfather *Don Pio* very well. He'd been a short dark Indian from *Oaxaca* and he'd been the one who'd taught our father so much about respecting life. My brother had hold of the rope on the hind leg of the steer and *Nicolás,* our top hand, had hold of the rope on the horns. My dad had his .22 Remington rifle in hand and all the knives were inside of the bucket with which we were going to catch the blood. After all, fresh, warm blood was a real delicacy on a good, properly done kill.

"I guess we're going to have to get him out of here and maybe kill him under the pepper tree," said my dad. "But we got to move him real easy, so don't make any sudden movements."

"Couldn't we just get some gas on a rag and rub his nose with it, so he can't smell anything?" asked my brother.

"That might work, but also it could just confuse him and . . . confused animals—just like confused people—can get a little wild, *mijito,"* said my father. "But what the hell, let's try it. *Mundo,* get some gas on a rag real *pronto."*

I quickly did as I was told. This animal was huge. We'd raised him since a calf. He was half Hereford and half Holstein. His face was all white, and his body was almost all black. I took the rag to our gas pump and rubbed it around the gas nozzle, then came back slowly towards the steer. If he panicked, he could really hurt me or tear up the place before they brought him under control.

"Relax yourself," said my father to me, seeing how scared I was. "If he smells any fear coming off you, we got trouble."

I took a big breath and blew out, trying my best to be fearless, but it was tough.

"Easy, big boy," I said to the huge animal. His head alone was bigger than me, and his horns had never been chopped and were long and razor-sharp. "This is just a rag with gas to confuse you," I said, "so we can kill you. Ooops! No, I didn't mean that," I quickly added, not wanting to panic the animal, and yet I also didn't want to lie to him, figuring if I was truthful, he'd feel better. "Look, we want you to smell this rag to confuse you so we can then—shit, *papa,* what am I supposed to say to an animal that we're going to kill?"

My brother *Chavaboy* and *Nicolás* burst out laughing.

"Here, give me that damn rag and you take the rifle and knives!" said my father.

I gave my dad the rag and he handed me his long-barrel .22 Remington and the blood bucket with the knives.

"Smell this, *cabrón,*" said my dad, putting the rag to the steer's nose. The huge animal took one sniff of the rag and leaped about three feet straight into the air, whirling about, farting a huge alfalfa-smelling fart, and went racing out of the old tractor barn. "HOLD HIM!" shouted my dad. "Work him over towards the pepper tree!"

Digging his heals into the ground, *Nicolás,* a real *vaquero* from the state of *Zacatecas,* held the huge animal pretty well, and my brother was doing well, too. But we could see that my brother was quickly tiring, so my father threw the rag away and took hold of the rope on the hind hoof with my brother. Pulling and holding, then letting the animal go when he was pointed the right direction, we finally got the steer down to the big old pepper tree.

"Tie him up," said my dad to *Nicolás,* "then go and get the hoist and chain from the tractor barn. This will give him time to calm down. And you," he said to me, "go get the grain and alfalfa."

Nicolás and I quickly did as told. My father stayed behind with my brother and the steer. My brother wasn't looking so good. I quickly ran to get the grain and alfalfa and bring it back to the steer while *Nicolás* climbed up in the rafters and got the tackle and chain off the main beam. Getting back to the pepper tree, I could see that my father looked pretty worried about my brother, and I could also see that my brother Joseph didn't want to be thought of as sick or weak, so he was holding on as strongly as he could. I had to fight back my tears. I loved my brother and I could tell he was hurting.

Finally, after about twenty minutes, the steer looked like he'd calmed down. He was starting to eat the grain and alfalfa and flick his tail to keep the flies off of himself. Talking gently to the big animal, my father now picked up his Remington and walked real carefully around the steer until he could get a clear shot at him between the eyes. And sure enough, one shot from that little .22 right between the eyes, just below the horns, and the huge beast went down like a ton of bricks, never knowing what had hit him.

Instantly, expertly, my dad moved in with one of his knives and cut the steer's throat just under the jaw, slicing the jugular so that the blood would now quickly pump out of his body, like nobody's business, and into the blood bucket. The bucket was almost full when the huge animal began to kick his first kicks of death—kicks so fast and wild and powerful that they actually yanked his whole body about and could have broken a grown man's leg if he wasn't well out of the way.

Suddenly, all the cats and dogs on the ranch were at our side, attracted by the smell of the kill. But they didn't dare move in on us, because they all knew that this kill belonged to us, the humans.

Nicolás climbed up into the big pepper tree, got the short thick chain over a stout limb, bolted the chain together, then hooked the chain and tackle to the short chain. As soon as the kicks of death had subsided, my dad moved in and cut slits in the steer's hind legs at the first main joint well above the hooves, then he hooked each hind leg into the end of a wagon bar, and started hoisting the steer up as we all began skinning him with our knives.

We all knew what we were doing, so we had the huge animal skinned in less than twenty minutes. We moved a big wheelbarrow under him between his front legs—because we'd hoisted him ass up— and my dad gutted him, then went through the intestines, looking for *las tripas de leche* that felt so warm and slippery to the hand. We also took the heart and liver, being very careful to not disturb the spleen or the piss bag. My mother would love the good job that we'd done, and she'd make blood pudding for us tonight, Mexican-style, in a real tasty *gesado* with plenty of onions and garlic—I could hardly wait!

That same afternoon, we dug a huge pit and lined the bottom with rocks, then that night, right after midnight, we started a huge fire with lots of good, solid wood that would take a long time in burning. This night I got to sleep outside with the *vaqueros* while they played guitars, drank *tequila,* watched the fire, and told stories of the old days when *hombres* were *hombres a todo dar* and they'd been doing this kind of killing and preparation since the beginning of time all over the whole world.

An hour before day break, my father came to check on the fire, which was now mostly coals, then he and *Nicolás* and a couple of other men went to the old ranch house with the huge kitchen where the women had been soaking the different meats all night in herbs and spices. My father and the men wheelbarrowed the meat of the steer and the two goats to the pit. Each piece of meat was wrapped in clean white sheets, and then wrapped in wet burlap and tied into bundles with good solid cotton cord. The three pigs we would cook aboveground in a great big copper kettle on an open fire to make *chicharrones.* We now lowered each of the huge bundles of meat into the burning hot pit with shovels and pitchforks. A couple bundles of corn and squash and sweet potatoes were put into the pit, too. Then iron pipes were placed across the pit and sheets of corrugated metal were laid crosswise over the pipes.

The men told me that in the old days in Mexico, they'd used green logs instead of iron pipes, and palm or banana leaves instead of metal sheets. Either way it didn't matter, they'd explained to me. What was important was to build a support so that the pit could then be covered with about five inches of dirt to hold in the heat. Every inch of the metal sheets was now carefully covered with dirt until not one little spirally pinch of smoke escaped.

Then it was done, and we now had ourselves an earthen oven, a way that people had been cooking, my dad told me, for millions of years all over the Mother Earth. The Father Sun was just rising on the distant horizon above the tree tops in beautiful colors of orange and red and gold when we finished. My father and the men passed around the last of the bottle of *tequila* and the jar of blood that had been especially prepared with lemon and spices.

"Now everyone get some shut-eye," said my dad. "Then we'll have *huevos ranchos* in a couple of hours, then get the rest of this feast ready. And remember, our guests will start getting here about noon and I don't want to see one of you *cabrones* drunk. This celebration is to last three days and nights, and we got to show these *gringos* that we, *los Mejicanos,* can have a celebration *a lo chingón* without getting drunk and ending up in a damn fight. Okay?"

The ranch hands all agreed and I went home to our new house with my dad. My whole head was full of all the different smells that I'd been experiencing in the last two days. All the animals slaughtered, all the innards separated, and all the food that was being prepared, *tortillas, frijoles, tripas de leche, chicharrones,* and that special combination of apple cobbler and *capirotada.* Smell was so important, and hearing was, too. The singing of the men around the huge fire was still ringing in my ears.

I went up the stairs of our new house and lay down in the room that I shared with my brother, but I couldn't sleep, I was so excited. My brother was down the hallway in our new bathroom and I could hear him coughing. For months now, my parents kept taking my brother to our local doctor who lived up the hill from us on California Street and had horses, but Dr. Hoskins kept saying that nothing was wrong with my brother, that he was just going through growing pains, and this was why he was always tired and not feeling well.

When I awoke I could hear people's loud, happy voices downstairs and I could see that my brother was lying down on his bed across the room from me and he was wide awake, but he didn't look so good.

"What is it, *José?*" I asked.

"I don't know," he said. I could see that he'd been crying.

"Do you want me to go and get *mama?*" I asked.

"No, I think I'll just stay up here and rest so *papa* and *mama* can have their celebration without worrying about me. You see, *Mundo,*" he added as he kept staring up at the ceiling, *"papa* and *mama* have worked so hard for this day that I don't want to ruin it for them."

I nodded. My brother Joseph, he was always so smart and good-hearted. I wasn't like him at all. I'd be yelling at the top of my lungs for *mama* if I didn't feel well, not giving a damn—I mean, a blessed—thing about their celebration.

"You really don't want me to call *mama?*" I asked again.

He turned and looked at me for the first time. "Promise me," he said, "that you'll say nothing, and if something happens to me, you'll always honor our parents." I swallowed. He was scaring me. "Look," he continued, "I've met a lot of rich kids and their parents at the Academy, and I'll tell you, there are no parents that I'm prouder to have as my father and mother than ours. They rose up with nothing, *Mundo,* except guts and faith and love. Do you understand?"

I shook my head. "No, I don't," I said.

He licked his lips. He seemed to be licking his lips all the time now. "You will," he said. "You just pay close attention, and you will."

He turned and began staring at the ceiling once again. I looked up at the ceiling, too, but I couldn't find anything worth looking at, so I got up to go.

"Remember," he said, "don't say anything to anyone. *Júrame.*"

"Okay, *te juro,*" I said.

"Good," he said, and he turned over, putting his face into his pillow. And I knew that he was crying again, but I didn't know what to do, so I just got dressed and went running down the stairs.

The front patio was full of people. They were all talking at the same time and looking real excited. I could see that most of them had arrived in cars and trucks, but I could also see that a few had come on horseback or buggy. We would have horse races in the afternoon and maybe even leaps of death, *pasos de muerte,* which meant jumping from a saddled horse to a bareback horse at a full run and riding the bareback horse to a standstill. The last time we'd done this, one of our *vaqueros* had been taken under a lemon tree by the bareback horse and almost cut to shreds by the thorns of our young lemon trees. But *Nicolás*, our foreman, who'd probably win the event, would have no trouble, because somehow he could mount a wild bareback horse and smooth-talk the horse with a gentle voice, and the horse would relax and trust him within seconds.

It was about noon when big George Lopez and Visteros came roaring up on their Harleys, grinning ear to ear like wild teenagers. My dad, who was now all dressed up in his half-*charro*, half-cowboy best, cussed them out, telling them that they were married men and shouldn't be risking their lives on those damn "murdercycles"! But George and Visteros had already had a quite a few, so they just laughed, and George told my dad, "How about you riding your horse through downtown Carlsbad drunker than a skunk!"

"But horses got brains!" yelled my dad back at George. "And my mare knew enough to get me home in one piece!"

"Well, Harleys got brains, too!" said George, laughing.

"All right," said my dad, "have it your way, but just no scaring the livestock this time, *cabrones!*"

"You got it, Sal! Now let's have a serious drink!" said George, grinning that drop-dead handsome grin of his.

Myself, I was already drinking lemonade and eating golden *chicharrones*, which were being cooked on an open fire in the great big copper kettle that had been brought in from *Guadalajara, Mexico*. The pork was a golden brown color and so delicious with a freshly made corn *tortilla*, a little lime, and *salsa*. By now there were already a couple of hundred people and the *mariachis* were playing. This was when a real good-looking young man from the *barrio de Carlos Malo* came by me, headed for the beer keg that was high overhead on a rack with blocks of ice. He had the biggest smile I'd ever seen.

"Hey, aren't you Salvador's son?" he asked me in Spanish.

"Yes, I am," I said in Spanish, too.

"Could you get me a couple of those *tacos de chicharrón?*"

"Sure, but come get them yourself," I said. "Everything is free."

"Yeah, that's because your father is the king."

"The king? The king of what?"

"Of all this territory."

"What territory?"

"From Orange County down to San Diego on the coast and inland from San Bernardino all the way to Escondido."

"What are you talking about?" I said. "Our ranch isn't that big."

He quickly fixed himself a couple of *tacos de chicharrón* with plenty of lime, *salsa,* and *cilantro*. "This ranch is nothing," he said, biting into his *taco*. "It's just a cover. Don't you know who your father really is? He's *el capón!*" he said with pride.

"Oh, you mean Al Capone, that guy in the movies."

"No, not Al Capone," he said laughing as he continued eating. "He's the real thing. He's *el capón,* the man who castrates, and he was so feared back in his day that people said that men's blood ran backwards to their heart, if they even thought of double-crossing him."

"My dad?"

"Hell, I was your age when it was said that he'd castrated *un hombre,* cooked his balls in *salsa verde,* and forced the man to eat his own *tanates.*"

"Oh, my God!" I said in disgust. "My dad did that?" It sounded awful.

"Hell," said the young man with admiration, "I personally saw him kill a man with his bare hands in the alley behind his poolhall when I was a kid!"

"You did." I said, and suddenly, I don't know how to explain it, but I could very clearly see that what this young man was saying to me could actually be true. How many times had I seen my father castrate livestock and then cook up their balls out in an open fire? A dozen times? No, more like a hundred!

My head exploded, my heart was pounding, and an icy chill went up and down my spine, as I looked at this young man who was devouring his *tacos de chicharrón* with such *gusto*. He was really good-looking, about eighteen years old, the same age as my sister Tencha, more or less, and it didn't seem to me like he meant to be speaking badly about my father. No, he was grinning, ear to ear, looking around at the huge place like he was in total admiration of *mi papa, el capón,* the castrater.

But oh, my God, to castrate a man and then cook up his *tanates* and force him to eat his own balls, that was pretty damn—yes, I mean pretty DAMN AWFUL! Then what was all this crap about my dad telling me to respect life, that to kill even a snail was being disrespectful of God?

I breathed, wondering if my brother Joseph knew this about our father. My heart continued to pound, and suddenly a dark cloud came overhead taking away the bright sunlight. I turned and saw *mi papa* over there talking to all these people, and he suddenly looked very different to me.

Flashes of the *barrio* came bursting to my mind's eye, and I now remembered like in a dream, my mother picking me up in her arms, waking up my brother and sister, and all of us running out the back door to hide behind the clothesline, as this huge awful monster came stumbling, bellowing to our home!

I could now remember all this so clearly. There were huge white sheets flapping on the clothesline like great white birds trying to take flight. My mother was crying and she held me close to her breasts. I could feel her terror. She was praying as fast as she could, begging God to please not let us be found. The tears were streaming down my *mama*'s face and getting me all wet. I couldn't figure out what was going on. Could this huge monster have been my very own dad? He kept bellowing something about, *"Mi FAMILIA! MI FAMILIA!* Where are you! I love you, *con todo mi corazón!"*

My brother Joseph wanted to go to the monster, but our mother held him fast. Then the bellow suddenly stopped, just like that. And my brother was the first one to go see the monster. And yes, the monster was our very own father, and he'd stopped bellowing because he'd passed out on the back porch into the awful-smelling puddle of his own puke.

Our mother told *José* to leave our dad alone, that sleeping in his own puke served him right. But *José* disobeyed our *mama*, went inside and got a blanket, and moved our dad's face out of the puke, being really careful not to wake him, and covered him up.

My mother wasn't crying anymore. No, now—with me still in her arms—she was looking up at the stars and thanking *Papito Dios* with all her heart. This, I now remember, was the first time in my life that I really looked up at the stars and saw how infinitely beautiful they were. I saw my mother had stopped crying and was talking to the stars. No wonder those big, ghostly white sheets kept trying to take flight—so they could be with the stars. The noise of the flapping sheets I'll never forget, for at that moment, I, too, began to flap my little arms, wanting to fly. The Stars were our True Home, I just knew it, here inside my heart and soul.

Then I snapped out of it.

I'd been like in a trance, and I saw that the housewarming celebration was going on all around me. This was when I saw that John Ford, whom my dad and I had met out by the front gates, was driving up in his baby-blue convertible. He had a woman with him who had wild red hair. I'd never seen hair like that on any human, horse, or dog. It stood up real tall, looking like flames of fire leaping off her head.

I quickly ran ahead to warn my dad, but I couldn't find him. There were hundreds and hundreds of people everywhere. I must've been off in my dreams for quite a while. I spotted Hans and Helen. They were talking with our town's mayor and several other city people. Jerry Bill, a close family friend, and Fred Noon, my parents' attorney, who were both from downtown San Diego, were walking up the driveway with a bunch of important-looking people. All these people had on real smooth, well-pressed clothes, clean shoes, and had a certain air about them. All of our neighbors and friends and *familia* had already arrived. My godparents, Vicente and Manuelita Arriza, had also arrived. Leo Meese and the Thills hadn't arrived yet. Jackie and Bert Lawrence, the most handsome young couple in all the world, were walking up. But only a few of our choice *amigos* from the *barrio de Carlos Malo* had been invited. My parents had made it known to everyone in the *barrio* that it was very important that there be no arguments between the *gringos* and *Mexicanos* on this day of their celebration.

My uncle Archie Freeman had been put in charge of security. He had several armed men riding the grounds on horseback, keeping watch over matters. Men on horseback, just like the knights of old, Archie had explained to my dad, could keep a large crowd in order pretty easily.

Finally, I spotted my father, but it was too late. John Ford and his red-haired woman were talking to Hans and Helen and to our Oceanside mayor. And now Hans turned around to introduce John to my father and mother, who'd just walked up. But when John Ford suddenly recognized my dad as the poor worker at the front gates, his whole mouth dropped open.

"WHAT!" shouted John, loud enough for everyone to hear. "YOU'RE THE OWNER!?! Why, you SON OF A BITCH!" he screamed, laughing. "I want MY FIVE DOLLARS BACK!"

"HELL, NO!" yelled my father, laughing, too. "You took it hook, line, and sinker, and that was the price for the story!"

"YOU CON ARTIST BASTARD!" said John Ford, still laughing. No one knew what John and my father were talking about. "Now, he's all dressed up like the *don*, himself," said John to everyone who was standing close by, "but three days ago, when he was working by the front gates with his son on a tractor, he was in the dirtiest old work clothes I've ever seen, and the bastard gets in my nice clean car, acting like he doesn't know any better and brushed all this dirt and cow manure over my nice, new—"

John didn't get to finish his words. Suddenly, out of nowhere, a lariat came down around John, roping him about the shoulders, yanking him up close.

"No, *Nicolás!*" yelled my father. "It's all right!"

"What the hell!" said John.

"Relax," said my dad, "it's okay, John. Everything's fine."

But *Nicolás* was hot, had the lariat dallied around the saddle horn, and was ready to give the horse his spurs and yank the *gringo* off his feet, and drag him through the orchard. My father loosened the rope around John's shoulders and slipped it off.

"Good throw," my father said to *Nicolás*. "Good job. But we got to be careful, and not over do it." Then turning to the crowd, he said, "Everything is okay. We just got a few horsemen working our celebration. Anyone who gets out of hand will be roped and dragged off, and then Archie will deal with you. This is a happy, peaceful occasion that Lupe and I have here. No guns will be fired up in the air! No fistfights or knife fights! This is a housewarming party!"

The woman with flaming red hair looked so excited that she could have popped. "Honey, I liked seeing you roped!" she said, snuggling up and down against John like a snake in mating season. "I want to learn how to do that!" Then turning to my dad, she said, "Then you must be the Al Capone of the West that my honey told me so much about?"

"No, not me," said my dad to the tall, beautiful, large-busted woman who just couldn't seem to stop her body from going up and down. "You got me mixed up with someone else. I'm just a plain little, law-abiding citizen like you and everyone else here."

"But honey, you told me that—"

"I'm Fred Noon, Salvador and Lupe's attorney," said Noon, coming up and introducing himself to John and the woman, whose name was Mary. "If you have any questions about Salvador and Lupe feel free to come to my office anytime."

John still couldn't stop grinning ear to ear, he was so happy.

"Damn you, Salvador!" he said to my father. "The only thing you didn't lie about was your wife. She is more beautiful than any movie star I've ever hired." And he bowed, taking our mother's hand and kissing her fingertips. "You are a queen from head to toe, Lupe," he said.

"You never say that to me!" snapped Mary, laughing.

"Of course not," said John, turning to her. "That's because you're my gorgeous wildflower."

"I'll settle for a wildflower any day of the week!" she said, grabbing him and kissing him hard on the mouth.

Myself, I was standing back and watching all this and feeling pretty confused. It seemed like everyone knew about my dad being *el capón* and . . . and . . . they liked it. This was really confusing to me. How could people like a man who put such fear into people's souls that their blood ran backwards from their heart? This made no sense. Did my dad then have monster frogs and lizards come up to him at night, too?

It was now time to uncover the pit. This was the highlight of the whole celebration. And it was at this very moment that Harry and Bernice, my father and mother's tailors, came driving up in their car. They were from Santa Ana and Harry had been the one who'd helped my father get my mother's diamond engagement ring for him when he'd asked for our mother's hand in marriage almost twenty years ago. Harry and his wife Bernice had the biggest, most beautifully wrapped housewarming present that I'd ever seen. It was huge! I could hardly wait to see what was in it.

"From Harry and me," said Bernice, handing the huge gift wrapped in silver paper to my mother, "for the two bravest, most loving young couple we ever met. Such struggles, and so much love. Harry and me saw these two fine people have during the Depression, and now look, a palace no less!"

Bernice was crying and my mother was crying and so was Helen Huelster and my aunts Maria and Luisa, but my *tía* Tota, she wasn't crying. In fact, she was poking Archie in the ribs and making a face at him like she was angry with jealousy.

The Father Sun was finally setting, and now everyone was gathered at the pit. For our father, this was the most holy time of the whole *fiesta*. He took a shovel from one of the workmen, and carefully, he started helping the other men remove the last of the dirt. Then he helped them take off the corrugated metal sheets. An aroma suddenly filled the air straight from Heaven!

People began *oooing* and *oh-oh-ing,* and then when the big bundles of meat and squash and corn and sweet potatoes were brought out of the

hole and placed on the long and heavy wooden table, people were salivating—the smell was so powerful!

"And now before we eat, my beautiful wife, Lupe, the treasure *de mi
corazón,* is going to make a toast," said our father.

"Yes," said my mother, stepping forward with grace and dignity,
"we're all going to be served champagne and give—"

"Served in real crystal, *a lo chingón!*" said my father, cutting in.

"Thank you, Salvador," said my mother, refusing to be ruffled by our
father's interruption. "We'll just wait for everyone to get their glass of
champagne, then we'll give our thanks."

I watched the people who my parents had hired for the day serve the
champagne to everyone. It was a long process. Even my sister Tencha
and our cousins Chemo, Loti, Andres, Vickie, Eva, Benjamin, and all
the others were served, too. Quite a few of my male cousins were still
overseas. One of them, my aunt Sophia's son, had actually made the national headlines as a war hero when he and his lieutenant had single-
handedly captured nearly a hundred Germans. But the newspaper
article gave the credit to the young lieutenant, when it had actually been
our cousin, a corporal, who'd come up with the idea and then handled
the *problema.*

"All right," said my mother. I could see that *mama* was very nervous,
but still she stepped forward with such calm elegance, as if she'd been
doing this all her life. "I'd like to start, as I said, by giving thanks to God.
But I've never spoken publicly like this before. I'm not educated, so
please pardon my English if I . . . I . . . I make any mistakes." She now
raised her glass of champagne to the Heavens. "Thank You, God, thank
You, Our Holy Creator, for giving my husband Salvador and I, Lupe, his
wife, the opportunity to build this great home for our *familia* and our
friends, new and old."

Just then, my brother came walking up and he signaled for me and
Tencha and our little sister Linda to come with him, up close to our parents. I was worried that our cousins would feel left out, but when I saw
our mother's whole face light up with such joy, as we four children

came up to her and our father, I knew that my brother had done the right thing once again.

"This is our *familia,*" said our mother to everyone. "Our two sons and our two daughters. Joseph, Tencha, Edmundo, and our newest child Linda."

I took my sister Tencha's hand. I could see that she was all embarrassed, too. My brother Joseph and my little sister Linda, on the other hand, didn't seem embarrassed at all. They were just glowing *con gusto* like our mother and father.

"As you go inside of the gates of our front patio," continued my mother, "you will see that we had a little glassed-in shelf built into the pillars, so we could house the statues of Joseph and Mary and Jesus. So, in the name of the Earthly Holy Family of Our Lord God, I invite everyone to now help us celebrate *nuestra casa* in peace and love and prosperity forever! Toast!" added our mother, touching her glass to my father's, then my brother and Tencha's glasses.

My little sister Linda and I had no glasses. I felt so left out. I could see that some people were actually wiping tears from their eyes as they lifted up their glasses to drink. Then my dad spoke up.

"This toast," he said in a loud voice, "that my wonderful wife Lupe just made is fine for her and I agree with her completely. Yes, *amor* and peace and prosperity are what we need here in this great nation of ours after that terrible Depression, and then this huge, long, awful World War Two.

"But, I'd also like to add that I, personally, didn't build this house just in honor of Joseph and Mary and Jesus. No, when we made plans to build this house, I immediately sent our architect to Hollywood to find how big Tom Mix's house was. Because when I first come to this country from Mexico, we see these Tom Mix movies in Arizona, with the *gringos* on the right side of the theater and the Mexicans and Blacks on the left side. And we see that no-good, fake son-of-a-bitch Tom Mix knock down five *Mexicanos* with one punch! And one Sunday in Douglas, Arizona—I'll never forget, I was just a kid—this big, handsome *Mexicano* from *Los Altos de Jalisco* got mad and jumped up on the stage in front of the movie and yelled, 'Come on, you *gringo* bastards! See if one of you can knock me down with one punch! And I'll give you the first punch free, *a lo chingón!'* And he ripped his shirt open and pounded his chest!

"And so—well, yes, of course, a fight got started. Two men were killed and ten more hospitalized. So I tell you, when we started to build this house, I told our architect, GO up to Hollywood and find out how big Tom Mix's house is, so we could build OUR *CASA* BIGGER AND BETTER! So I now say to all of you that I didn't have this house built just for peace and love, but to also tell every DAMN HUMAN BEING ON ALL THE EARTH that here in Oceanside, California, stands *UN MEXICANO DE LOS BUENOS CON SUS TANATES* IN HAND, free to work or fight with both hands, whichever way the DEVIL WANTS TO PAINT IT! And this is MY TOAST *A LO CHINGÓN! SALUD!"*

SHOUTS ERUPTED!

A GUN WAS FIRED—once, twice, three times!

And all the *Mexicanos* present went wild *CON GRITOS DE GUSTO!*

Instantly, the man who'd fired his revolver was roped and taken off through the orchard. The *mariachi* started playing. People drank their champagne, were served more, and they began to line up to get their food.

At the main table, the mayor and his wife sat down with our mother, Hans and Helen, Noon and his wife, and half a dozen other people. Then here came my father carrying the huge platter covered with a white cloth. It smelled of Heaven! He set it down in front of our mother and the mayor's wife, pulling off the cloth.

This was an honor, the highlight of the whole *barbacoa*. The skinned out steer's head with the horns, eyes bulging, and the mouth open with the tongue sticking out, surrounded with *chiles* and other colorful spices. It was a delicious, mouthwatering, beautiful sight. But the mayor's wife let out a SCREAMING screech of pure terror, and fell over backwards along with her chair.

I couldn't figure out what was going on. Hell, my father hadn't even taken a *tortilla* to scoop the brains out of the huge skull to pass around as a kind of Mexican caviar to be eaten with the champagne, yet.

Quickly, my mother and Helen took the screaming mayor's wife inside the house to calm her down.

My father, feeling pretty insulted by her outburst, said, "To hell with it! Let's switch from champagne to *TEQUILA!*"

The mayor, trying his best to keep calm and go along with everything, accepted a big shot of *tequila* from *Los Altos de Jalisco* from where comes all the best *tequila!*

And our great big housewarming celebration now became a real *pachanga!*

CHAPTER **eleven**

Two weeks after the party, a car full of *hombres* drove up to our new house late one afternoon, honking their horn. They were from our old *barrio* of Carlsbad. I knew two of them. It was obvious that they were drunk and mad as hell. They just kept honking until we finally came out of the house.

"YOU'RE A TURNCOAT, *GRINGO*-LOVING, son-of-a-bitch *cabrón!*" one of them shouted at my father as we came outside to see what it was that they wanted. "You've forsaken your own people and gone over to the *gringos* like a cheap *puta*-whore!"

"You're all just drunk," said my father, "so I'm going to ignore what you say, but you better turn your car around and leave before you get me mad."

"*¿Y que? Eh,* what will you do?" said the biggest man in the front passenger's seat, opening his door. "You're not so tough anymore!" The man was so big that when he got out, the springs of the car rose up. "I alone can kick your ass, Salvador!"

My brother and I were behind our dad inside the gates of the front patio of our new home. The statue of the Blessed Mother was in the pillar to the left of us and the statue of Saint Joseph holding baby Jesus in his arms was to the right of us. I didn't know what to do. There were four of them, and the one who'd got out of the car looked so much bigger and stronger and younger than our father that I was ready to pee in my pants. I thought of going inside to get a gun. But not my BB gun. I'd get my dad's .12 gauge shotgun, but I didn't know how to shoot it. And if I did, it would probably knock me over backwards.

"Don't move," said my brother to me under his breath, as if reading my thoughts. *"Papa* knows what he's doing."

I swallowed, hoping to God that my brother knew what he was talking about. Because, my God, I had no idea what our poor, old *papa* could do against this giant.

"I said," said my father, as if he had all the time in the world, "we're *amigos,* and you're all just drunk, so I'll ignore these insults that you tell me here at my home, in front of my two sons, but just this one time, so you better now turn around and get out of here. *Mijito!"* he said, turning to me, "get me a bottle of whiskey so that these fine gentlemen can enjoy a drink on their drive home."

"Órale!" said the one who was driving. "Now you're talking, *compa!"*

The giant grinned and licked his lips, salivating at the thought of another drink. Quickly, I ran into the house. But in our bar, I discovered that most of our liquor was gone. So I just grabbed a big bottle of something almost full, I didn't know what, and came racing back out. I was really terrified for our father.

"Here," I said to my dad, handing him the big bottle.

My father took the bottle and started to hand it to the huge man who was standing in front of him. But swaying back and forth on his feet, the gigantic man took a closer look at the bottle that my father was handing him, and he ROARED!

"THIS IS SHIT, Salvador! That ain't a man's drink! It's that SWEET GREEN SHIT FOR WOMEN! Who do you think we are, you son-of-a-bitch, *cabrón!"*

Without a moment's hesitation, our father leaped at him, smashing the bottle across his face! The huge man dropped to the blacktop like our thousand-pound steer had dropped under the pepper tree when the little .22 bullet hit him between the eyes. Our dad then broke the liquor bottle across the open door of their car, holding the bottle out like a weapon towards them, with all its broken, ugly razor sharp edges. Suddenly, my brother had a snub-nose .38 in his hand. It was our father's gun. Joseph had been inside in bed when the car had first driven up. But he didn't need to do anything with the snub-nose. Three of our *va-*

queros now came running up and they had a crowbar, a lariat, and *Emilio,* our new foreman, had my dad's shotgun.

"Put him in his car," said my father to our cowboys, "and make sure these damn fools never come through our gates again! Why do you think I didn't invite you, *cabrones?* It's because of this, you stupid sons of bitches!"

"It's not over, Salvador!" yelled the driver.

"Of course not, as long as you breathe, you'll be a damn fool who can't open his eyes and see what's going on!"

They drove off, tires spinning and fish-tailing up the driveway. I stood there transfixed, suddenly realizing that this was the same kind of action I'd seen in the *barrio* as a child behind our poolhall—I'd just forgotten. That good-looking young man's story about my dad being *"el capón"* came flashing back to my mind. My God, it had to be true. It all had to be true. There just wasn't any doubt about it in my mind anymore.

I started crying. I'd caused this whole thing to happen. If I'd brought out a bottle of men's drink instead of women's, they would have driven off happy, and I would've never endangered *mi familia.*

When my father saw me crying, he came over to me. My brother had already put his arm around me.

"It's okay, *mijito,*" said my dad to me. "You don't have to be scared anymore."

"I'm not scared," I said, shoving my brother's arm away. "I'm mad! I brought out the wrong liquor, and that's why all this . . . this happened!"

The workmen laughed, saying that I looked cute being so angry. But my brother and father didn't laugh.

"Look, *mijito,*" said my dad, "no matter what kind of liquor you would have brought out, it would've still ended up in a fight. That's what they came looking for—a fight. It couldn't be no other way."

Emilio was nodding in agreement with my father. "That's why I brought the shotgun," he said. "I could smell their want of fight from way across the orchard."

"You can smell somebody wanting to fight from across an orchard?" I said, realizing that was one hell of a long way.

"Sure. Why not? Bulls can smell a cow in heat when the wind is right, five miles away," said *Emilio*. "Anger, fear, sex, these are strong smells."

"Exactly," said my father, "just as every married man must know, or he's gonna get hit in his *tanates* with things from his wife that he has no idea where they're coming from. Smell, she is very important, *mijito*."

"Is that why you cooked that man's *tanates* in *salsa verde* before you forced him to eat them, so he could smell them?" I asked. Ever since the housewarming celebration, I just hadn't been able to get those two stories out of my head. One, my dad castrating a man and cooking up his balls in *salsa verde;* then, the other, killing a man in the alley behind our poolhall in *Carlos Malo* with his bare hands. Also, every time I remembered those two stories, I'd remember that night with the white sheets flapping on the clothesline.

My dad was grinning. "Who told you that one?"

I was shocked. My dad was grinning like he truly enjoyed the memory of that awful story. "A guy at the house celebration," I said.

"Yes, you're right. I wanted that man to smell his own *tanates* cooking, because I needed to teach that bastard a lesson he'd never forget for the rest of all his *pinche* life!" said my father with sudden anger. But then he breathed, and breathed again, calming down. "You know," he said to me, "I think the time has come where we maybe need to have a little talk." And saying this, he took hold of me by the shoulders and turned me around. "You see, *mijito*, now that you are coming into manhood, you need to know what it is exactly that a man is and what it is that they do. Back home, at this time, was when my brother *José*, the Great, taught me how to shoot a gun, so no man, no matter how big, could ever walk on my shadow again without my permission.

"*José*," he said, turning to my brother, "give me that pistol you brought out, and go in and get the .22 rifle and a bunch of bullets so we can teach your little brother how to shoot. My father," said my dad, turning back to me, "he never taught me nothing, but hate and anger and how abusive men can be with all their power. It was my mother and her side of the *familia, los Indios,* who taught me why it is that God made a man's balls so sensitive and easy to hurt."

My father had his huge hand on my left shoulder as we walked around our new home. "You see, to be a good *hombre a las todas* is to be—just like a man's balls—soft and tender inside your heart, and yes, easy to be hurt. This is why all good men need to be raised like a woman for the first seven years of their life. Those men, who came in that car all drunk, they just don't know nothing about being soft and tender and honest, deep inside."

I nodded. This really made a whole lot of sense. "What men do know this, *papa?*" I asked.

"Not many anymore," he said. "Nowadays, even most women are raised up to only admire the male way of thinking. The respect for women was lost," he added, "when people started saying that God is only male. You see, back at one time, all over the world, two stories of Creation were known. One for the women, with a Female God, so they could teach the young girls, and the other one for the men with a Male God, so they could teach the young boys."

"You mean, there were two Bibles back in Eden?" I asked.

"Yes, of course. That only makes sense. No man and woman could ever agree on just one story."

"Well, then, why do we now only have one Bible?" I asked.

My father grinned. "You guess."

"Because," I said, "of men?"

"Exactly," he said. "And how did you figure that out?"

"Because *Eva* doesn't blame Adam. It's Adam who blames *Eva,*" I said.

"Very good thinking," said my father, grinning. "You got it. That's why it's so important to raise boys as girls for the first few years of their lives—so that, then, they'll at least have a little understanding of the—"

"Then was Jesus also raised like a woman for the first seven years of His life?" I asked.

"Of course. That's why He was so strong and yet tender. All men back when the Earth was young and the Stars spoke and people knew how to listen, were raised like women for the first seven years of their life. In fact, back then, there were even two tongues. One for women and one for men, and the men who weren't raised like women for the first years of their life didn't even know how to speak to a woman when they grew up."

"That makes a lot of sense," I said. "Because I've heard some of our ranch hands say that there's no talking to a woman."

My father laughed. "And they're right," he said. "There is no talking to a woman in a man's language."

My brother Joseph came up with the .22 rifle that my father used for killing the pigs, goats, and steer for the housewarming celebration and a couple of boxes of ammunition.

"*Papa,*" I now asked, "then, like Jesus, do you forgive those guys who came up in the car, because they didn't really know what they were doing?"

"Forgiving, that's a tough one," said my dad, taking the rifle and beginning to load it. "Sober, they're actually good men, especially the driver. So, sober—then, yes, I guess, I do forgive them."

"But drunk you don't?"

"No, drunk I don't. I'm still not as big of *corazón* as Jesus," he added.

I nodded. Things were now beginning to make a whole lot of sense to me once again. Then maybe my dad wasn't such a monster after all. *Emilio* and two of our ranch hands came up.

"Here," said my dad, handing me the pistol. "Take a good feel of this .38 Special. It's loaded," he said. "Always remember this, guns are always loaded, whether they're loaded or not, so always keep your finger off the trigger and keep the barrel pointed that way, towards our target area on the hillside, where it's safe in case the gun goes off. I've seen many a man shot and some killed with what people thought were unloaded guns. Get it?"

"Yes," I said, "I get it."

"Good," he answered me.

We walked all the way around behind our new home and halfway down the hillside. We were just up the hill from the tullies where the deer and fox lived and the red-shoulder blackbirds came into roost each night by the tens of thousands. My dad took his time telling me all about weapons, and especially about this snub-nosed .38 Smith and Wesson, and how it compared to other handguns, like the .44 and .45.

"And some men spend their whole lives looking for the best gun, just like other men spend their lives looking for the best wife," he said, "but these men are really just fools in the end, and cowards, too. Because, *mijito,* when it's all said and done, it's never the gun or horse or wife or car or truck that does a good man in or saves his life. No, it's all here between the *hombre's* own two legs and how well he knows how to handle that gun, horse, wife, car, or truck. Give me any gun, and I'll do good no matter how the Devil paints it for me. Why? Because I know what I'm doing, and . . . I've had a lot of practice."

"Is that the same for women?"

"What?"

"That they need a lot of practice, too?" I said.

Emilio and the men smiled.

"Well, no," said my dad. "We men like to think that it's only us, the men, who need to practice."

"This makes no sense," I said. "Why shouldn't women practice, too, so they'll know how to handle a gun, horse, husband, car, or truck, too?"

Hearing this, *Emilio* and the other workmen burst out laughing, as if I'd asked the most ridiculous question in all of the world. I got mad and swung around at them with the .38 in hand before I realized what I'd done. They all jumped back.

"NO!" screamed my dad, grabbing me and turning me back around. "Watch what you're doing! I told you never to point a gun at anyone! And you," he said to *Emilio* and our two *vaqueros,* "if you can't keep quiet, then get back to the barn!"

They immediately calmed down, putting on serious faces. My brother never laughed.

"That's a good question," my father said to me. "That's why they got nervous and laughed. They weren't poking fun at you. You see, *mijito*, the truth is that for our women, we men like them not to be experienced when we marry them. But in time, yes, we do want them to learn how to do things well," said my dad.

"But why don't men want women experienced when we first marry them?" I asked.

"Because, well," said my dad, glancing around and seeing that *Emilio* and the other men were trying their best to hold back their laughter. "Because, *mijito*, most men, when they're thinking of marriage, prefer to do their own training of a woman, just as they do with their own horses. And also, men don't end up with babies and women do, and so each man is afraid to think that it's another man's child that he's raising."

"But why," I said. "You just told me that when it's all said and done, that it doesn't matter which gun or horse or truck or wife, so why should it matter whose child?"

Now *Emilio* and the two men couldn't hold back their laughter, no matter how much they tried. Even my own brother was smiling.

"Because, well, *es mucho más fácil criar que quitar mañas*," said my father.

"Exactly," said *Emilio*, "*es mucho más fácil poner rienda que enderezar una mujer mula.*"

"Thank you," said my father to *Emilio*. "Are you beginning to understand, *mijo*? It's easier to raise than remove bad habits. It's easier to put on rein than to try to straighten out a stubborn, mule-headed woman."

I shook my head. "But all this still makes no sense," I said. "You've always told me that women are ten times smarter and tougher than men. So then, why don't women get the experience, and then train us, the men?"

Well, I must've said the most crazy-*loco* thing in all the world, because now one of our *vaqueros* was laughing so hard that he fell down on the ground, holding his stomach. They were all in wild hysterics.

"What's so funny," I yelled. "I like riding on a horse that's experienced and already well trained, so why not with a woman?"

The howling laughter of the men continued, no matter how hard my dad tried to stare them down. Now, even my own brother was laughing.

"You know," said my father, "I think that we've talked enough for one day. Remember, you're only eight years old. In time, you'll understand. Now let's get back to the easy part of being a man, and teach you how we shoot a gun."

"You mean shooting guns is easier than knowing how to be with women? That makes no sense, either. I get along real easy with my *mama* and my sisters, and they're women."

"Fine. I'm glad to hear this," said *mi papa*. "And that's the way it should be. Now, pay close attention," he said, taking the .38 snub-nose away from me and unloading the weapon. He handed it back to me, and turned me about facing the hillside and had me dry-fire the revolver without bullets at the hillside. The .38 felt big and heavy and when the hammer was back, the trigger was real light.

"Now watch," he said, reloading the gun. "You go first *Chavaboy.*"

Taking the snub-nosed revolver, my brother shot at the hillside, not quite hitting the little rock that he was shooting at. Man, the gun was loud! Even the horses way out in the valley lifted up their heads to look towards us. Myself, I'd had to cover up my ears real tight with my hands. My brother shot five times.

"I've never been very good with handguns," said Joseph. "I'm much better with a rifle. *Papa*, he's good with either one."

"Not that good," said our dad, taking the .38 Special and reloading it. "I just get lucky sometimes." He quickly took aim and started firing, blowing that little rock to SMITHEREENS!

Boy, was I ever impressed, and so were my brother and *Emilio* and the other workmen. Then it was my turn. I was nervous as hell. I'd never fired a real gun before.

"Use both of your hands and square yourself, facing forward," said my father. "You'll have to get a lot bigger to handle a .38 or .45 with one hand. But always remember, if you ever have to face another man, you use one hand and turn sideways, so this way, you're a smaller target and you can move and slide."

I did as told, squaring myself to the target. And I'd shot a lot with my BB gun. But my God, when I took aim with this big black, heavy snub-nose and squeezed back on this trigger, the weapon suddenly leaped up in my hand like a snake, almost breaking my wrists, and sent EXPLODING FIRE out the barrel!

And the sound! My ears were ringing like nobody's business! Everyone was laughing. I guess I was staring cross-eyed at what I'd just shot. I'd missed the little rock that I'd aimed at by five feet. Handguns were really very different from rifles. All afternoon we spent shooting. The workers got to shoot, too. With the .22 rifle my brother was really good. But our dad, my God, he could hit the center of a nickel every time.

"In the *barrio, papa* would throw quarters into the air and hit them," said my brother with a big smile. "That's how he stopped the fights between the Marines and the *vatos.*"

"You'd stop fights, *papa?*" I asked.

"Sure. Of course. I was the peacemaker of the *barrio,*" said my dad. "Just as Archie was the law and peacemaker before me. So when the Marines came to the *barrio* asking for trouble, I didn't insult them. They were just kids, so I'd take them behind the poolhall and teach them how to shoot. After the war, two of them came back saying that I'd saved their lives with what I teach them."

"Yes!" I said excitedly. "Now I remember! You'd shoot at quarters on the tree by *mamagrande*'s house, wouldn't you?"

"Yes, that's right," said my dad.

"Then you did good things in the *barrio*, too?" I asked.

"Sure. Yeah, of course," said my dad.

"But I was told that you killed a man with your bare hands in the alley behind our poolhall," I said.

"Who told you that one? Eh, the same guy who said I castrated a man and forced him to eat his own *tanates?*" I nodded. "Listen to me closely," said my father, "very closely. Yes, I did castrate a pig and forced a man to eat the pig's balls. Then I pulled down that man's pants, asking him if he preferred to eat his own balls cooked in *salsa verde* or *salsa colorada*. But he never stopped screaming, he was so terrified, so I never had to castrate him. And when I let him go, he went back to Los Angeles, telling his people to never come back to *Carlos Malo*. Get it, *mijito*? I got done what I needed to get done to protect my territory. All bears do this. All lions do this. And that man in the alley, yes, I beat him, and then I put him in the trunk of my car and drove him out to that big eucalyptus forest behind Carlsbad so he could walk it off before he caused any more trouble. The next day, he left town back to Mexico. The people of the *barrio* never saw him again, so some started saying that I'd beat him to death and took his body out towards Vista to bury it."

"So then, why haven't you told people that you didn't kill that man and that you really didn't castrate that other one?"

"What? And RUIN my reputation!" said my father, laughing. "Hell, no! I worked too hard to get my reputation! And fear is a damn good thing, *mijito*, when the other guy has it and you don't!"

I shook my head. All this really didn't make any sense to me. But I could see that the workers loved my dad's explanation. I guessed that maybe there was still a lot for me to learn, even though I was already eight years old.

The Father Sun was setting when we started up the hill. The shooting was done. *Emilio* and the *vaqueros* had left to go and take care of the stock. It was just my brother and father and me, walking back towards the house.

"*Papa*, those men in the car," I asked, "if you knew that it was going to end up in a fight anyway, because you could smell it, then why did you tell me to go get some whiskey?"

"Because, *mijito*, maybe it could've helped," said my father. "And also, because it gave me a little more time to think, caused the driver to remember that we're friends and to call me a '*compa*,' and gave *Emilio* time to come with help."

"Just wait," I said, "then you mean that those guys are friends of ours?"

"Well, of course," said my dad, laughing. "No enemy would ever come up to a home angry because they didn't get invited to a celebration. You just remember, *mijito*, whenever your wife comes screaming at you that she's going to kill you because you forgot her birthday, that what she's really saying is 'I love you, I depend on you for my life and love, and so this is why I hate you and want to kill you!' It's always with our wife, our best friends, and our relatives that we end up having most of our troubles in life, *mijito*.

"A man's enemies are never his biggest *problema*, that's all just a bunch of bullshit lies in the stupid, no-good fake movies, like Tom Mix always used to make, making us *los Mexicanos* into bad people. *Mi mama* always used to say there are no bad people on Earth, once we've learned how to open your eyes and really see."

"But, *papa*," I said, "Superman, he's a superhero just like Batman, and they're always out chasing bad people."

"Those two guys aren't married, are they?" asked *mi papa*.

"No, they aren't," I said.

"Tell me, where would Superman be if he was married and his wife was mad at him for being late for dinner or he forgot her birthday or anniversary because he was out chasing bad guys?"

My eyes went huge. "My God, he'd be in deep sssh—I mean, trouble, *papa*."

My brother started laughing, but this wasn't a laughing matter for me. This was really important. Yeah, sure, I'd always known that Superman and Batman weren't really real, that they were cartoons, but the stories of what they did, I'd always thought were good and true and was what we, the men, were supposed to do if we wanted to be heroes in our own lives.

"Then, *papa*," I said, "are you saying that Superman and Batman are fake and cowards because they aren't married."

"You tell me," said my dad.

The sudden fear that went shooting all through me made me want to stop thinking about all this. I'd never questioned the Bible or Superman or Batman before. It was bad enough when I'd questioned Santa Claus, because I'd begun to think that there was no way that he could get around the whole world delivering presents in one night. I'd told my school friend What-A-King that reindeer weren't that fast, even if they could fly. I figured that it had to be our parents who gave us our Christmas presents. What-A-King went bananas and started crying, then hit me with his rollerskates across the head, almost killing me. I was beginning to find out that what we'd been raised to think was no small laughing matter.

As I went up the hill that night to our home with my brother and father, my whole head was swimming. All this stuff about going into manhood seemed a hell of a lot harder than having been raised like a woman, where everything had just made such damn—I mean, blessed—good horse sense.

CHAPTER **twelve**

A screeching ambulance rushed my brother Joseph to Scripps Hospital in La Jolla! My parents had finally taken our brother to another doctor— a young man recently out of the Navy. Dr. Pace just took one look at Joseph, saw how yellow the whites of his eyes were, and immediately had him rushed to Scripps Hospital.

After several days of tests, Dr. Pace explained to my parents that my brother's liver was ruined. My dad wanted to know how this terrible thing could have happened. Dr. Pace said he didn't know. My dad told him that was bullshit! That he could see it in his eyes that he wasn't telling them everything. Dr. Pace added that if Joseph had just been hospitalized a couple of months earlier, they could have put him on juices, particularly cranberry juice, and his situation would've never gotten this far. Coming home that night, my father was so mad that he wanted to drive up California Street to kill Dr. Hoskins with his bare hands. It was only my mother's screaming and pleading that finally stopped our dad from killing the doctor who'd been attending to Joseph all this time.

Specialists were brought in from across the country. Every week, my brother was taken back and forth from our home to Scripps in La Jolla. And for me, school continued, and now all of my friends, including What-A-King, were in the fourth grade, but I was still in the third. My brother was hurting and the magic went out of my marble playing. A lot of the kids started beating me, even the smaller kids who were in the third grade with me. Once more I began to draw stars every chance I got. Five-pointed ones and six-pointed ones, then I'd color in the stars with green and blue and little touches of red and yellow.

One day my parents' arguing got so bad—while my brother was down at Scripps—that I went upstairs to draw my stars so I could disappear. My little sister Linda followed me and found me drawing stars on a big piece of cardboard on the floor outside on our balcony, which overlooked all of our property. She told me that she wanted to draw, too. So I showed her how I'd recently learned to draw stars with a ruler so I could get the lines real straight, then I'd color them in, and it would make me feel really good.

All that afternoon, my little sister drew stars with me, and all this time we could hear our parents arguing downstairs. Our mother was crying and saying that she was sure that God was punishing her and our father for something that they'd done, and so she was asking God to please take her and not our brother.

"Please, dear *Papito*, our son Joseph is innocent," she was saying to God. "No child should be held responsible for their parents' sins."

"Lupe, why do you talk like this," said our father. "God doesn't punish children for their parents' mistakes. Come on, Lupe, we just need to find the right doctor. Dr. Pace says that there's still hope."

"Salvador, this is what Original Sin is all about. God passing the sins of the parents on to the children."

"But didn't Jesus come to get rid of all those *pendejadas?*"

"Well, yes, that's why I'm praying."

"Well, then, pray for *Chavaboy,*" said our dad, "to get well, not for you to die instead of him. I love you, Lupe," added our father, "and the kids need you. Don't pray for shit like that!"

"But what if God has already decided to take Joseph. Then I'd rather go instead of him," said our mother.

"Lupe, DAMNIT!" bellowed our dad. "You will not be praying like this in our home!"

"You can't tell me how to pray or not pray, Salvador, here in our home or anywhere else," said our mother. "This is between me and God!"

"Then I'll KICK GOD'S ASS if he listens to you when you pray like this!" he yelled. "I'm not Adam! I'm not afraid of God or His threats of hell! I'M *UN MEXICANO DE LOS BUENOS,* and nobody is leaving, do you hear me! Nobody is getting punished! Shit, we just moved into our new home, for God's sake!"

"Salvador, I will not permit you to talk to God like this!"

"Oh, I can't tell you how to pray, but you can tell me how to talk to God, eh?"

"But you talk vulgar!"

"And you talk stupid! God don't punish nobody! That's all just a bunch of bullshit from the Church to scare us! And if anyone is gonna be punished, I'll take it! Not you, not *Chavaboy!* Me! Here, CHEST FIRST! I'm not running out on my *familia* like that chickenshit Adam who blamed his wife *Eva.*

"YOU HEAR ME, GOD! I'm talking to You, man to man, *a lo MACHO CHINGÓN!* It's between YOU AND ME!"

"Salvador, you can't be ordering God around! Talk nice!"

"BULLSHIT! Nice never gets us nowhere, and you know it, especially with God!"

Hearing all this, I continued drawing stars as fast as I could, then coloring them, but now I wasn't jumping inside them so they could take me away. No, I could now see very clearly that I needed to stay here on Earth to help *mi familia.*

My parents' shouting continued. My mother was crying and so was my father, but for some reason, it didn't frighten me anymore. No, now those little quiet vibrations had once more began humming behind my left ear. I continued drawing stars, and the humming behind my left ear began to spread across the back of my head. Then, suddenly, I felt safe here on Earth, too. I didn't have to go away. *Papito Dios* was also here with me.

This was when I realized what this humming really was: God was massaging the back of my head with His fingertips. I smiled, feeling so good, and just kept drawing stars as fast as I could while my parents kept yelling and arguing downstairs.

Later that same night, I went outside after dinner so I could be alone and look up at the stars in the heavens. I just didn't want to hear my parents arguing anymore. Hell, I agreed with my father—no one from our *familia* was going to be taken away or punished. But if someone had to be taken away, then I'd be the one to be taken and punished.

After all, it only made sense. My brother was smart and was needed here. And my father and mother, they were our parents and so they were both needed, too. But me, I was stupid, so I wasn't needed. Hell, every day at school, I was getting stupider. Now most of the new third graders could also read better than me. I was going to flunk the third grade again. I began to cry. The way I saw it, I was probably going to be kept in the third grade for the rest of my life. I'd be old and have all white hair and no teeth and still be in the third grade. And the kids, they kept telling me that this was because I was a Mexican, and Mexicans were only good for washing dishes or working in the fields like stupid animals.

As I was standing by myself in the front patio and looking up at the stars, something real strange happened. Suddenly, this little goldfish leaped out of the water in the fountain beside me, and I swear that he smiled at me.

I laughed, wiping the tears out of my eyes. Now I remembered that when I'd been little and my *mamagrande* had rocked me to sleep by the water fountain at our home in Carlsbad, I'd also seen goldfish smile at me.

"How are you doing, *mijito?*" asked my father. He'd just come outside and joined me in the front patio.

"Good, *papa*," I said, lying. The tears were still running down my face.

He was smoking his cigar. My mother was inside praying at her little altar to the Virgin Mary in the hallway outside our parents' bedroom. My little sister Linda was praying with her. I felt sorry for my little sister. I sure as hell didn't like it when my mother caught me and had me kneel down with her and do all that praying to her candles and altar.

Carefully, my father turned his big cigar about in his mouth as he continued smoking. I kept looking down at the fish in our fountain in the front patio. I wondered if fish ever slept, or did they just keep swimming around and around all day and night.

"Do fish ever sleep, *papa?*" I asked.

"Do what?"

"Do fish ever sleep? And if they do, do they lie down at the bottom of the fountain to sleep?"

"Hell, I don't know," said my dad. "I guess they do. Why wouldn't they sleep?"

"That's what I was thinking, and I bet that they dream, too. But at school they tell us that only humans are evolved enough to dream or think."

And saying this, I don't know why, but my eyes again filled with tears.

"What is it, *mijito?*"

"Oh, I don't know," I said, realizing that my parents had so much to worry about with my brother Joseph that I didn't want to add to their worries. No, I had to be strong and keep still. "I'm okay."

"Is it school?"

I nodded.

"What is it?"

"Well, the kids, they laugh at me, *papa*. And the teacher does, too. You see, the other day I told them that animals dream and that they know how to think, too. Because, *papa,* you and I have both seen our horses jerk in their sleep, and also I've seen them think and figure things out, like how to open gates and get into the grain room. I've also seen dogs toss in their sleep, and whimper like they're having bad dreams." The tears wouldn't stop. "They just kept laughing at me, *papa,*" I added. "Especially when I told them, as you've told me, that there are some horses that are so smart that they can out-think most humans."

I didn't say what this one kid had said. "Sure, your horses can out-think you Mexicans, because you're all so stupid, you flunked the third grade!"

"So school still isn't going so good for you, eh?" asked my father.

"No, *papa*," I said, wiping my tears. "I'm not learning to read any better than last year."

He put both of his hands in his front pockets and rocked back and forth on the balls of his boots. My father's boots were beautiful. He always had to have them custom-made or had to have a zipper put in on one side, because his arch was so high that he couldn't slip his foot into a regular size slip-on cowboy boot.

"I don't know what to tell you, *mijito*," he said, still smoking. "I don't know how to read so good myself."

"*Papa*," I said, "did you used to sometimes come home from the poolhall in Carlsbad so drunk that you were stumbling and yelling and *mama* got so scared that she'd take me and Joseph and Tencha to hide out by the clothesline?"

"Who told you that one?"

"No one. I remember it. The white sheets would flap in the wind like great big birds trying to fly."

"I don't know," said my dad. "But it could be, I guess, true. Some nights it really got bad in the poolhall. A liquor store is a much better business. Here, the people buy their liquor and go home to drink and pick their fights. They don't all try to take their shit out on the bartender."

"Why do you sell liquor, if it's so bad?" I asked. "Maybe this is why *mama* thinks that God is punishing you two."

My dad took a big breath. "God isn't punishing anybody," he said to me. "It's us, humans, who punish each other left and right, and have screwed up this whole wonderful world of ours. Take us, the people, off this *planeta*, *mijito*, and the whole world has not one *problema*. Nothing wrong with liquor," he added. "It's here in each man's heart, what he brings to his drink that turns everything into something terrible. Remember the old Mexican saying, 'In the first half of the bottle we find God, then in the second half is where we find the Devil.' "

"Oh," I said, "then we should throw out all the bottles after you drink the first half."

My father laughed. "No, it just means that too much of anything is never a good thing. So yes, that was probably me coming home drunk after a terrible night of breaking up fights. You see, there was no law in the *barrio*, except me, so every son of a bitch who had an argument with his wife or boss would come and try to take it out on me. Being a cop, especially without a badge or the jailhouse to back you, is hell! No wonder Archie sold me the place so damn cheap. He was tired. And after ten years, with no days off, I began to get ugly, too. Your mother was probably right to get you kids and run and hide. Hell, it must've been like the whole damn Revolution for her all over again."

"Why didn't you just quit?" I asked.

"Because, *mijito*, a married man with children never quits his job. No matter what, a married man with children bites the bullet and just keeps going day in and day out to work. These are the real heroes of life, men who bring home the bacon. Do you know what it means to be a *burro macho?*"

"No, I don't," I said.

"Well, then listen, and listen closely, because this might just be the most important thing you'll even learn about becoming a man."

"I'm listening."

"You see, back in the old days all supplies were transported by horses with wagons in the valley and by a string of *burros* up in the mountains. And back then, one day's travel was about twenty to twenty-five miles, the same distance as the padres built the Missions here in California. And a *burro macho* means, that when the *burro* driver comes to his regular watering place at the end of a long day and he finds no water or food, his *burro machos* don't break down. No, these fine, strong animals just pull in their guts and go right on to the next water hole—another twenty or twenty-five miles—without arching their back or trying to dump their load. You see, the other animals, who aren't *burro machos*, don't want to go on, so they arch their backs and try to dump their load, or worse, they start twisting and make welts on their backs, preferring to get sick and break down, than go on. It's no accident when a workman breaks his leg or his shovel on the job. Accidents happen when a man doesn't have the *tanates* to pull in his guts and go on with the heart of a *burro macho*.

"You see, over and over I've seen men at the end of a tough job get tired, start complaining, get impatient and break their leg or shovel because they'd quit in their hearts, but wanted to pretend in their heads that they're strong and had worked hard to their last. The head is weak, *mijito*. Always remember this, it is in our hearts that we men are strong. So now, do you see why the highest compliment that any man can give to another workman is to call him a *burro macho*? A *burro macho* never quits or breaks down, no matter what!"

I nodded. "I got it, *papa*," I said. "Then a man's real power comes from the heart, just as it does for a woman?"

"Exactly! Now you got it!"

"All right, I'm glad that I now got it, but still, shouldn't a man be very careful to decide at what he's going to work at so hard?" I asked. "I mean, couldn't it still be . . . that all those bad things that happened at the poolhall are the things that *mama* was talking about God maybe punishing Joseph, instead of you and—"

My father EXPLODED, screaming so loud that he scared the pee out of me. "I WILL LISTEN TO NONE OF THIS BULLSHIT!" he roared. "*Chavaboy* is going to BE ALL RIGHT! We don't need God to punish us for NOTHING! Don't you understand! We do a DAMN GOOD ENOUGH JOB OF PUNISHING OURSELVES ALREADY, especially if we're good people like your mother!

"Hell, the bastards of the world do horrible, monstrous things, then live their whole lives happy, all the way to their old age! It's the saints that suffer, the good people like your mother that got this whole damn thing ass-backwards, ever since we got our asses kicked out of *Paraíso!*

"WE DON'T KNOW SHIT about living life with God anymore, I tell you! My mother, *Margarita*, nothing but a bag of skinny Indian bones, she didn't let herself get fooled into all this hocus-pocus bullshit of the Church! She told us all, ever since I can remember, 'They came to teach us with a Holy Book that only talks about life after people lost their way of living in the Grace of God. Why listen to that?' she'd say to us. 'I want to learn from a book that tells us about the millions of years that people lived in the Graces of *Papito* back in the Holy Garden. To listen to anything less is an insult to our Immortal Souls!'

"God is Love, *mijito,* NOT REVENGE!" my dad now told me with tears coming to his eyes. "You hear me, my mother taught us that God is *Amor,* and Jesus said, 'Forgive them for they don't know' on the cross as they were pounding nails into him. Well, I forgive them all, and I forgive myself first of all, and that's how it is, and all the rest is *caca de un toro muy viejo, y PENDEJO!*"

Tears came to my eyes. "Then *mama* really isn't going to have to leave us and go to Heaven instead of Joseph?"

Hearing this, my father looked at me for a long time. "Is this what you've been thinking?" he asked. I nodded, with tears streaming down my face. "No, *mijito,* your *mama* isn't going to leave us."

"Really?"

"Really."

"Good, because if anyone has to go, then . . . I . . . I want it to be me, *papa.* I'm the one who isn't smart."

"Oh, my God," said my father, taking me in his arms. "So this is why you were crying when I came out here. The things we put into children's heads, damnit!" His whole chest expanded against me. "You are a very good, brave *hombrecito a las todas,*" he said. "I love you *con todo mi corazón, mijito,* and nobody is leaving. We're all staying here together and working things out. We're a *familia,*" he added. "Always remember that, we are *una familia,* and a *familia* always finds a way for things to work out, *con el favor de Dios.*"

And so my father held me close and I cried and cried, feeling so much better. I'd thought that our mother was going to have to leave us for sure, and I didn't want her going, or my brother, either.

That night, I dreamed of Stars and being up in Heaven with *Papito Dios.* I was flying through the Heavens, and when I came to this Golden Eye, I shot through the Eye and here was my old dog Sam and my two *mamagrandes.* Sam came up and started licking my face and Doña Guadalupe and Doña Margarita had medicine bags with them, and they walked me through the Heavens, teaching me how to pick Stars and put them into my own medicine bag.

And it was so easy, all I had to do was just walk up and take a Star in my hand, any Star, and then rub it between my palms until it was as warm as my hands. Then I smelled of it, and if it smelled all sweet to me, I was to put the Star into my medicine bag.

Soon my whole medicine bag was glowing from all the Stars I'd picked. This was so much fun! Each Star I'd picked felt so warm and good and full of Love, just as *mi papa* had told me, that *Papito Dios* was *Amor!*

CHAPTER thirteen

My dad was right! Nobody was going to have to leave *nuestra familia!* My mother's prayers had been answered! My big brother was feeling much better. In fact, he felt so strong that he was back home with us, eating *menudo,* which a doctor from Mexico City had recommended. And now, after a week of homemade *menudo*—tripe soup cooked Mexican style—he wanted to go horseback riding.

I ran to the stables to get *Emilio* to saddle up Midnight Duke for my brother. Joseph had his own horse, a beautiful sorrel-red stallion named Blue Eyes, a gorgeous quarterhorse that we kept at Jimmy Williams's place over by Escondido, but he never rode Blue Eyes. He always preferred to ride Midnight Duke, who was a gelding.

Myself, I saddled up my old mare Caroline, and then my brother and I went riding out of the corrals and down the hill towards the railroad tracks. I was all excited being with my brother, so I suggested that we ride inland, over towards Crouch Street where I'd recently found a herd of deer. But my brother told me that he wanted to go the other way, down the canyon to the ocean.

"I want to ride along the seashore," he said. "I don't know why, but I've been having dreams of riding Duke in the surf. Do you know the best way to get passed the marshlands?" he asked.

"Yeah, of course," I said. "I do it all the time."

"Then lead the way," said my brother.

So I turned my old mare Caroline and started down the valley, going through the marshy part of our *rancho grande* on our way to the ocean. I knew every step of these marshes, but I could see that my brother didn't quite know how to maneuver through the little waterways and keep up with me. Also, he just wasn't as quick and agile with his horse as I was with mine.

We could now hear the ocean in the distance. We were where the marsh grasses grew so thick, that they covered up some of the deep crevices the water had dug into the Earth after a hard rain. I wasn't paying too much attention to my brother's progress anymore. No, I was seeing new places where Shep and I could go hunting. Lately, Shep had started teaching me how to hunt ducks. I loved duck hunting with Shep. In fact, we'd gotten so good at hunting ducks, that they were now becoming my favorite thing to eat, cooked with fresh oranges and honey just like they did in a Chinese restaurant. My brother called out to me.

"*Mundo!*" he called. "How did you get over there?"

I turned around in my saddle and saw my brother had come down the inner side of the last little waterway which I'd crossed about fifty yards back, and now he couldn't get to the flat where I was.

"You can't cross there!" I yelled back at him. We were about forty yards apart. "The water is too muddy and deep! You'll have to go back and cross where I crossed!" I continued yelling.

"Where is that?"

"Way back there!" I shouted, pointing.

He turned Midnight Duke around and started back up the valley. I didn't turn my horse to go and help him. I was watching the profusion of wildlife down here at the west end of our *rancho grande*. My God, I'd never realized there were all these little fish in these waterways down here, close to the sea. I wondered if the bigger fish came in with the high tide and ate these little fish.

"Do I cross here?" shouted my brother.

I turned and looked, and I saw that *José* was facing a big clump of thick marsh grass, and so yes, it could be safe to cross, but on the other hand, it could be one of those bottomless breaks that was just all covered up with grasses.

"Yes, I think so!" I yelled back at him.

Then I saw Midnight Duke gather his feet under himself as my brother reined him towards the clump of grass. Instantly, I knew that I was totally wrong and this was a dangerous place to cross, and Duke knew it, too, and so this was why the big black gelding was gathering up his hooves to leap.

My big brother didn't seem to realize why the horse was gathering his feet under himself, so I screamed, "NO! DON'T DO IT THERE!" but it was too late, and Duke leaped like a great black panther, trying to do what he was being asked to do.

My brother, not expecting this, terrific leap, was jerked back in the saddle. I saw his whole face go white with pain. He hadn't moved forward to help the horse with his leap so Duke didn't make it across the over grown, invisible waterway, and his hind legs went down into the hole of muck and water.

My brother was fighting with all his strength to stay in the saddle. The horse was now leaping and lunging, trying to get his hind quarters out of the ditch. The clump of grass was being trashed down into the hole. Finally, with heroic strength, Duke and my brother made it to good solid ground.

All this had taken just seconds. And by this time, I'd raced my own horse across the flat and I was at my brother's side, I could see that Joseph's eyes were welling up with tears. He was in so much pain that he couldn't even speak.

"I'M SORRY!" I said. "OH, MY GOD, I'M SORRY! I'll get you home!"

It seemed to take us forever to make it back home. He couldn't run his horse. It hurt him too much. And he didn't want me to run ahead and get help, because he didn't know his way around the marshes. I felt like such a damn fool! If only I'd waited for my brother, staying by his side. I'd been a stupid, selfish asshole!

As soon as we got home, my brother was put to bed and Dr. Pace was called. When Dr. Pace arrived, he asked my brother what had happened.

"I told you not to exert yourself," he added.

"It's not his fault," I said to the doctor. "I was the one who knew the trail."

"It's nobody's fault," said my brother. "My horse just got stuck in the marsh and began lunging. It was all I could do to just stay in the saddle," he said, trying to laugh like a real cowboy.

Dr. Pace, having been raised on a ranch in Arizona, laughed and continued examining him. "If you don't feel the pain subside by morning," he said, "we'll have to take you back to Scripps."

I ran out of the room feeling terrible, and that night my mother didn't have to ask me to join her at her little altar to pray with her. I knew that it was all my fault that my brother Joseph was hurting. I prayed and prayed that night as I'd never prayed before, but it didn't help. In the morning, my brother Joseph was still in so much pain that they had to rush him to the hospital. And all that day at school, I felt so terrible that it was hard for me to hear anything the teacher was saying.

She was a substitute teacher. We were told our regular teacher was sick. It was in the late afternoon when this teacher looked down her sheet and called my name to stand up and read. I didn't hear her—I was so busy drawing stars.

"Young man," she said to me, coming down the aisle between the seats, "I see that you haven't read in some time. I'd like you to now stand up and read for me."

I glanced around. Everyone was staring at me. My heart started pounding a million miles an hour. "No," I finally said.

She looked at me. She was shocked. She was a lot younger than our regular teacher. "What did you say to me?" asked the woman teacher.

The whole room was silent, wanting to see what I'd say or do next. None of these kids were really my friends. They were all a whole year younger than me. "I said, no," I said.

"You said no to me?" she said, starting to breathe faster.

"Yes, I said no to you," I said.

"Well, we'll see about that!" she said, grabbing me by the ear to pull me out of my chair. "You will stand up and read when I call on you!"

Some of the kids started laughing. Others began making catcalls. She was strong and pulled me out of my chair, no matter how much I held on to my desk. Then she put my book in my hands.

She turned my book to the right page. "Read!" she commanded.

My eyes were crying and all the kids were watching me. I couldn't make heads or tails of all the words on the page that just kept jumping and running all over the place. She slapped me on the head.

"Pay attention, and start reading!" she said once again.

Still I couldn't read, and when she went to slap me again, this time, I don't know why, but I'd had enough. So I dropped the book and whirled about and grabbed her hand. Growling like a dog, I bit her as hard as I could.

She screamed out in pain! I loved it! And I bolted out of the classroom, running as fast as I could. I was going to get on my bike and pedal home and never come back to school ever again! I'd take my dad's .38 Special and my BB gun and get my horse and really run away this time.

But I never made it to the bicycle rack, which was under the pepper trees. Before I could, the playground teacher caught me. He was a big man and he grabbed me, twisted both of my arms behind my back, and marched me to the principal's office. In the office was the substitute teacher, holding her hand and telling the principal that I bit her like a rabid animal. I thought she said "rabbit" and started laughing, because I'd bit her a whole lot harder than rabbits can bite. The principal didn't think it was funny so he and the man, who'd brought me, held me by my arms, as the substitute teacher slapped me again and again, calling me names. My eyes cried and cried, but I swear, that I didn't cry inside of me. No, I took my punishment just like Ramón . . . and yes, Jesus had done, never saying a word.

Getting home that afternoon, I didn't know what to do. I still felt like running away. But another part of me felt like getting my father's .38 Special and a couple of sticks of dynamite and going back to school and shooting all the kids who'd laughed at me, and then blow that damn principal and substitute teacher to smithereens. Especially that playground man who'd twisted my arms until I thought he'd broken them. But before I could decide what to do, run away or go to school, my parents told me to get in the car with them, that we were all going to go to downtown Oceanside to see the priest about having some masses said for my brother.

Getting to Saint Mary's Star of the Sea in downtown Oceanside was really strange. I'd never been inside of a church except on Sundays, when it was full of hundreds of people. Going inside, I dipped my fingertips into the holy water at the entrance, made the sign of the cross over myself as I'd' been taught to do, and followed my parents down the left side aisle towards the apse. The main aisle at the middle of the church was only for weddings, funerals, or show-offs, I'd always been told.

A tall priest was waiting for my parents near a picture of the Holy Virgin. He greeted my mother and father by their first names, then he suggested that my little sister and I stay out here and my parents go into the back room to confer with him alone. My mother asked me and my little sister if we'd be all right. We both said yes, and sat down in a pew and watched our father and mother disappear into the back of the church along with the priest.

The inside of the church was real cool and semidark and smelled kind of good, in a funny way. Glancing around, I realized this church was really tall and long and beautiful. When we came here to mass on Sundays, we weren't ever allowed to just look around. No, we had to keep our eyes on the priest at all times and pretend we were really paying attention to all the stuff that he kept saying and saying. It was so boring.

This time, glancing around, I could see that the sunlight streaming in through the windows made long, golden columns of light across the inside of the church. These columns looked so beautiful. They kind of reminded me of the sunlight that I'd seen coming down through that Golden Eye that I'd dreamed about when I'd been traveling through the

Heavens. I sat there with my little sister in our pew and the silence was so peaceful that suddenly, just like that, I began to feel that humming begin behind my left ear once again.

I smiled. Maybe *Papito Dios* still loved me. Maybe God hadn't left me even after what I'd done to my brother. Tears came to my eyes. Maybe my dad was right and God didn't want revenge and to always be punishing us. Maybe *Papito* really was *Amor,* and the *problema* was just us, ourselves, since we'd lost the Garden, kept punishing and fighting with each other.

I glanced at my little sister Linda and saw that she'd stretched out on the pew beside me and she'd gone to sleep. I sat there glancing all around at the tall, magnificent church—the beautifully carved beams, the gorgeous stained-glass windows—and the humming behind my left ear grew and grew.

I wiped the tears from my eyes, and I was now kind of glad that I hadn't gone back to school and killed all those people. After all, a good man always ate what he killed, and I couldn't have cooked up and eaten all those kids and the principal and that substitute teacher. I laughed. This was really funny. I would have had to dig a big hole and make a huge fire to barbecue that many people.

I was laughing and glancing around when I suddenly noticed all these stars all over the church—in the wood, the pictures, the glass windows, and even up on the altar on the clothing of Jesus and Mary and Joseph. Stars were everywhere, and they were all smiling.

Suddenly, I knew here in my heart that God had forgiven me. "Thank You, *Papito,*" I said, making the sign of the cross over myself. "Thank You, very much."

I guess that I, too, must've gone to sleep, because when my parents came out and woke my sister and me up, I'd been gone for a long time, laughing and dancing through the Heavens with Sam and my two *mamagrandes!*

As we left the church, our parents decided that we'd go out for Chinese food. We drove in our car one block west and half a block south, next to the movie theater. Archie and our *tía Tota* met us at the restaurant. Archie was loud and happy, telling my parents that everything was going to work out, not to worry, that Joseph was going to be fine, that he and George Lopez had gone down to Scripps today and donated blood for him. Hearing this, I could tell that my father and mother felt a lot better, until our *tía Tota* opened her mouth and said, "This is why I never had any kids! I just can't stand all the suffering that I've seen my sisters go through, first at birth and then when one of their children dies."

"Our son isn't dying!" said my dad, coming out of his chair.

"Salvador," said our mother, grabbing hold of him. "She really didn't mean it that way. You know how Carlota just talks."

"It's not just talk," said our *tía*. "It's the—"

"Here, have a shrimp!" said Archie, shoving a great big shrimp into our *tía's* open mouth, shell and all.

"Archie!" she yelled, gasping. "You know I hate fish!"

My father burst out laughing.

"Shrimp ain't fish," said Archie. "Just eat and shut up!"

"It tastes awful! Worse than lobster!" yelled our *tía*.

"Have another!" said Archie, going to put another shrimp into her open mouth.

"Don't be putting things into my mouth!" yelled our aunt.

Laughing, feeling much better, my dad now reached across the table to take my rice, which I'd been saving for last.

"NO!" I screamed at the top of my lungs. "That's my dessert!"

Everyone in the restaurant turned to look at us. My dad put my rice back down, looking very sheepish. "I thought you didn't want it," he said.

"It's my favorite," I said. "That's why I was saving it for last!"

"This kid is definitely weird," said Archie. "I always eat my favorite things first. Eh, honey baby?" he said to our *tía Tota*, pinching her.

When we left the restaurant it was already dark. We were all very happy. My father couldn't stop laughing. "Did you see how everyone looked at me when he screamed about his rice?" he said. "I bet that they were saying, look at that dirty old man, stealing the food from that little kid."

"Yeah," said Archie, "this one has been a hell-raiser ever since he was born. I'll never forget when he pissed on you, Sal, and he was still in diapers!"

This story I'd been hearing for as long as I could remember. It had happened back in the *barrio de Carlos Malo,* when my dad had been building a doghouse for my dog Sam and I'd wanted to help, so I took my dad's hammer and started hitting on boards, too. My dad couldn't find his hammer, so when he saw me with it across the yard, he called me a *cerotito,* meaning little turd, and he came across the yard, took the hammer away from me, and playfully whacked me on the butt.

I remember very well that I'd gotten mad and come over and grabbed the hammer away from him, calling him a little turd in Spanish, too, as best I could. And he said, "You don't talk to your father like that," and he really whacked me good this time.

The rest of the day I'd stalked my dad, just like our cat stalks a mouse. I'd followed him around all afternoon. And when he finally went inside to take a nap before going to his second shift at the poolhall next door, I sneaked in, climbed up on the coffee table, pulled my diaper down, pissed all over him as he slept, getting my golden stream of urine to go into his open, snoring mouth. When he sat up coughing and gagging and spitting, I took off out the back door and got under the house with my dog Sam.

And no matter how much my dad yelled at me to come out from under the house so he could whip me, I refused. Finally, he had to go to work, and at the poolhall, he told everyone how I'd pissed on him. I became the hit of the *barrio,* the only person who'd pissed on Salvador and gotten away with it.

Late that night, Uncle Archie had come over all the way from Oceanside, just to find out if the story was true. My dad had gone next door and taken me out of bed where I was sleeping with my mother. He brought me back to the poolhall where he put me on top of the bar top to show everyone how I'd pissed on him. I refused, feeling all embarrassed, but my dad had then pulled my pants down, completely exposing me to everyone, so in anger I pissed on him again. My dad went wild, wanting to really hit me good, but Archie pulled my dad off of me, telling him that this was the greatest thing he'd ever seen, because, if the truth be known, there wasn't a man alive who hadn't dreamed of pissing on his old man!

That night in the poolhall, my uncle Archie had hugged me close, laughing and laughing, and gave me a silver dollar—that I still have—and told my dad that he'd arrest his ass if he ever heard that he'd laid one hand on me for pissing on him.

And now this night in front of the Chinese restaurant, I saw Uncle Archie take my father in his great big arms and give him a strong *abrazo*, chest to chest.

"We've been through so much, *amigo*," Archie told my dad, "Probation, jail, bullets, and we'll get through this one, too. Look, I got an idea," added Archie. "Why don't you and Lupe come with me and Carlota to Las Vegas. We're going to see a few shows and do a little gambling. Hell, you've done all you can. You've seen the priest, paid for some masses for Joseph, and you got the blood that you needed from me and George. Come on, you and Lupe need to get out of town and breathe a little."

"Maybe you're right," said my dad, "and we'll take George and his wife, Vera, too."

"Who's paying?" asked Archie.

"Hell, I'll pay," said my dad. "Good God, Archie, all my life my biggest problem was to get enough money so I could just eat and keep the wolves away. Now Lupe and me got more money than we'll ever need, and good God, I still got *problemas*. Maybe even the most painful that I've ever had. Joseph, he's such a good, decent young man *a las todas*, and he's suffering so much, damnit!"

My dad had tears running down his face. "Yep," said Uncle Archie, "we get out of one frying pan of trouble only to find ourselves in another pan that's even hotter."

• • •

Our parents were off to Las Vegas less than an hour when my sister Tencha and her girlfriends decided to throw a big party. But they hadn't gotten permission from our parents, so they had to swear everyone to secrecy. My brother Joseph was home once again, looking pretty good after the transfusions of blood he'd received. Our big, handsome cousin Chemo, who was more or less my sister Tencha's age, was visiting. He was our *tía Maria's* son, and *tía Maria* was the older sister of my mother and *tía Tota*. Chemo was a senior in high school and a local football hero. He told my sister Tencha and her girlfriends that if they supplied the girls, he'd supply the guys. My brother Joseph wasn't saying anything. He was just lying down in a lounge chair in the front patio, taking it really easy, like he'd been told.

When my turn came to be sworn to secrecy and they asked me to promise to not tell our parents, I said, "Sure, but it will cost you fifty cents each for me to promise."

"Fifty cents each!" yelled my sister's friend, Virginia. "That's blackmail! Where did you learn that?"

"From Chemo," I said.

"You learned blackmail from Chemo?" they all asked, glancing at our big, gorgeous cousin.

"Yes," I said. "He paid me a whole dollar to not tell anyone that I found him in the barn with—"

"That's enough!" said my cousin, laughing and grabbing me. "He's only joking."

"No, I'm not," I said. "Don't you remember, you paid me to not tell anyone that you were in the barn with—"

Quickly he covered my mouth with his hand, grinning at my sister and her girlfriends. They were all looking at him.

"So who were you in the barn with?" asked Joan, another of my sister Tencha's friends. She liked Chemo a lot. You could tell by the way she was always smiling at him.

"No one," said Chemo, his face turning red. "The kid is just kidding—get it? Kid, kidding!"

But no one laughed. They just kept looking at him.

"All right," I said, getting his hand off my mouth, "then it will be a dollar each if you don't hurry up and pay me!"

"Look," said Chemo, taking me aside and whispering, "I gave you that damn dollar to keep quiet, but we can't just keep paying for everything we want to do. Blackmail isn't a joke, *Mundo*. It's terrible, especially when you do it to your own relatives."

"But why, Chemo? It sounds pretty good to me. You big, old guys want a party, so you get your party, and I want a few dollars so I can buy myself a pint of strawberry ice cream and a new bow-and-arrow set. My old bow is too weak, now that I'm bigger."

"Look, we're not going to pay you a cent," said Chemo, "and that's that!" he added in a strong voice.

We were in the back patio next to the waterfall with the tall ferns and flowers. Birds were singing in the aviary. It was a beautiful day. And my brother Joseph really did look a lot better, but I could also see that he didn't really care about the party one way or the other. It was Chemo and my sister Tencha and her girlfriends who wanted the party. Back then our local Oceanside High School also took in all the students from Carlsbad, Vista, Bonsall, and San Clemente, so Oceanside was the center for all the teenagers in the whole North County of San Diego. Also, with Chemo being drop-dead handsome and a star football player, Tencha and her girlfriends really wanted him involved in their party. I had them over a barrel.

"All right, don't pay," I said, "but then I'll just tell *mama* and *papa* about your party as soon as they get back from Las Vegas."

"I'll throw you in the garbage can and ship you out to sea with the rest of our city's garbage!" shouted Chemo, getting angry.

"Hey, you're not going to throw him in a garbage can," said my sister Tencha, coming over and stepping in front of Chemo.

"But Tencha, he shouldn't be trying get rich off of us!"

"Why not?" I said. "It's you guys who are hiding? And *papa* always says that where there's fear, there's money to be made."

"But not off your own family!" insisted Chemo.

"All right," I said, "then I'll only charge you twenty-five cents each, the price of an arrow."

My brother Joseph, who was sitting over to the side, never said a thing. He just kept watching. Finally, a deal was made, and that afternoon each person who came to the party put a quarter in the little basket that I'd put on a table in the front patio. I quickly told my sister Tencha about the ones who didn't pay and then she and her girlfriends would either bankroll the person, if they wanted them to stay, or ask them to leave. I loved it! I was getting rich just like our church in downtown Oceanside, with the tall, beautiful ceilings, that collected money on Sundays and sold masses during the week.

When I saw the blonde arrive—the one who'd been with my cousin Chemo in the barn on the haystack—I acted like I didn't recognize her, so I wouldn't embarrass her, and just asked her for her quarter, too.

"Is this a house tax?" she said, sarcastically.

Instantly, I liked the sound of this much better than blackmail, so I said, "Yes, it's a house tax for the party."

She paid and walked across the patio to be with Chemo. I could now see why our cousin was so embarrassed about anyone finding out what they'd been doing in the haystack in the barn. She wasn't that big or strong-looking, and yet when I'd climbed up the ladder and taken a look, I'd seen that she was on top of our poor cousin. And they'd both looked like they were in terrible pain, but still she'd been able to hold him down and keep riding him like a cowboy rode a bucking *bronco*.

That night at the party, I climbed up into the apricot tree outside of our back patio with my pint of strawberry ice cream. I'd never had a whole pint to myself. All my life, I'd only been allowed to have one scoop at a time. I loved blackmail—I mean, taxes. Because with my own money, I could buy anything I wanted anytime I wanted. And best of all, I was still young and just getting started, so who could tell where

I'd be in just a few weeks or even months if I kept my eyes open and learned all I could about black—I mean, taxes.

I was up in the apricot tree, eating my big pint of Carnation strawberry ice cream, my favorite, and keeping watch on the party. I'd collected money for twenty-six people, and from up here in the tree, I figured I could keep track of anyone new who came into the party, so I could collect their money, too.

The party was just beginning to get kind of good and wild, and I was truly enjoying myself, leaning back on a branch of the tree, eating my ice cream in big luscious spoonfuls, when I suddenly felt dizzy. I dropped the carton of ice cream and I fell out of the tree just like the squirrels that Shep got to fall off of the fence posts.

The next thing I knew, I was crying and crying and my sister Tencha was at my side, trying to help me. But my cousin Chemo and his blonde girlfriend weren't trying to help me at all. No, they were laughing and saying that I deserved what I'd gotten because I was a greedy little pig.

The rage that came out of me, I don't know where it came from, but instantly, I stopped crying. I didn't care if my body was broken or not, and screamed. "YOU WILL NOT LAUGH AT ME! Do you hear me? You do, AND I'LL TELL on both of YOU!" I shouted.

"YOU LITTLE SHIT!" yelled Chemo, lunging at me. "We had a deal! I'll break your neck!"

The music stopped. My brother Joseph came up. "You won't be breaking anyone's neck, Chemo," he said in a soft, firm tone of voice.

"Joseph is right," said Tencha, also coming up alongside my brother Joseph to protect me. "That kind of talk isn't right, and you know it, Chemo! We're family!"

"But he gets away with murder!" shouted Chemo.

"Not murder," said my brother, "just a quarter each, and that's a small price to pay so we can have our party in peace."

"That's easy for you to say," said Chemo, his whole face and body twisting.

Then I saw it. I could smell it coming off of my big, strong cousin. Chemo was scared to death of anyone finding out about him and that blonde on the haystack. I was shocked. He was so big and brave and strong out on the football field, knocking people left and right when he ran with the ball, and yet, here it was fear that I'd been smelling coming off him all day long. I guessed he was deathly afraid of people finding out that a girl had been on top of him instead of him, the rooster, being on top of her.

"Chemo," I whispered to him, "I'll never tell anyone about you being on the bottom."

"What?" said Chemo.

"You know, on the hay, when she beat you up."

Hearing this, Chemo started laughing and laughing. "So that's what you saw?"

"Yeah, but I won't tell anyone that you lost."

"Good," he said. "Thank you! That's great!"

The music started once again and everyone went back to the party. My cousin Chemo seemed happy now. I guess he just didn't realize that women were ten times stronger than men, and so it was okay if they won.

I looked down at the ground, saw my crumbled, soggy carton of Carnation strawberry ice cream, and realized it was all gone. All this work to get a pint of ice cream, and I'd still ended up with probably no more than a scoop.

I was scared. We were going too fast. It was the day after the party and I was with Chemo and my sister Tencha. We were in Joan's parents' car. They were all three up in the front seat and I was in the back. Chemo was driving and we'd gone up California Street and turned right. We were barreling down the hill on a curvy dirt road leading to the duck lagoon between Oceanside and Carlsbad. They were laughing and having a grand time.

"Slow down," I said. "You're going too fast!"

"Shut up, you little twirp!" shouted Chemo back at me. "I know how to drive!"

Normally, I didn't get scared when someone drove fast. In fact, I liked it and thought it was fun. But this time it was different, and I just knew that something terrible was going to happen to us if I couldn't get our cousin Chemo to slow down.

"Chemo," I said, "please, I'm not joking! You got to slow down!"

"Shove it!" he said, giving the car even more gas.

Then I saw it in a flash, a great big, old red truck was coming up the road from the bottom of the hill with a huge load of lumber, taking up most of the steep, little road.

"OKAY, CHEMO!" I yelled. "This is your last chance! I swear, if you don't obey me and slow down right now, then I'm going to open my door and jump out and you can all get killed by yourselves around the next corner, hitting a big red truck!"

"Sounds good to me," said Chemo, laughing. "Jump! I don't see any truck, red, green, or purple," he said, refusing to slow down.

Now what could I do? I'd opened my big mouth and he'd taken my bluff. In my mind's eye, I could see the big truck changing gears and coming up the hill, so clearly inside of my head. The man driving the truck had on a beat-up old cowboy hat and he was trying to light a cigarette as he drove, and when he looked up, I recognized him. It was Bert Lawrence, the man who shoed our horses for us, and Bert, a big handsome man, was married to Jackie, the most beautiful woman in all of Oceanside, next to my mother Lupe and George Lopez's wife Vera.

I got hold of the door handle, and when I felt Chemo slow down a little bit as he took the next curve, I opened the door and jumped.

The dirt road came rushing up into my face with lightning speed, and I hit the stone-hard gravel road with a scream of pain and went rolling into a ditch full of rocks on the side of the road!

Tencha SCREAMED!

Chemo BRAKED!

The red truck SWERVED!

Tencha's girlfriend was in wild hysterics, and her parents' car barely missed the big red truck.

All this I saw happen like in a slow dream, as I hit the dirt and felt my face and right shoulder get cut to pieces on the gravel.

My head snapped back against one of the rocks in the ditch and I was gone!

I wasn't here anymore!

I was dead.

Stars were all about me. Beautiful stars.

Then, I realized that I wasn't just dead. I was also outside of my body, about twelve feet up in the air. I'd seen my own body go rolling off the road and my head snap against a rock.

From above, I now watched my sister Tencha leap out of the car as Chemo came to a quick tire-skidding stop, gravel spraying everywhere. She was screaming and calling for me, her "Shrimpito!"

Tencha was the only one who ever called me "Shrimpito." "PLEASE, LORD GOD!" she screamed as she came back up the steep brown gravel road. "DON'T LET HIM DIE!"

I was really enjoying watching everything from up there. I'd never realized that roads and streets looked so smooth and wide from above.

"I'll kill the little shit if he isn't dead!" Chemo was shouting as he got out of the car, all red-faced with rage.

"No, you won't!" shouted Tencha at Chemo. "He was right, Chemo! You were going too fast! We would've all gotten killed by that truck if he hadn't jumped out and caused you to brake!"

"That's a lie! I know how to drive!"

"But how did he know about the red truck?" asked Tencha's girl-friend Joan.

Hearing this girl's question, I then saw it so clearly. Even when I'd been alive, I hadn't just been in the backseat of the car as we'd been speeding down the curvy gravel road; no, I'd also already been outside of my body—up above—seeing myself down below and everything else just like I was now doing.

I laughed, and watched my sister Tencha and Chemo and Joan come running up the steep hill, but to me it looked like they were going in slow motion. Then they went down into the ditch where my dead body lay still against the rocks.

"SHRIMPITO!" Tencha yelled.

"OH, MY GOD!" shouted Chemo. "The damn fool! I thought he was joking! No one is really stupid enough to jump out of a speeding car!"

I could now see that Chemo loved me, too. He was just all confused and scared.

"Is he dead?" asked Joan.

Hearing Joan's question, I suddenly realized that this was my decision. I could go either way. I could be dead or I could be alive. Living or dying was really all up to us.

I took a deep breath. I didn't know which way to go. A part of me really liked being dead and way up here above everyone and everything. But also, I could see how much my death was hurting my sister Tencha who loved me so much.

"I want to live," I finally said to my Other Self, and saying this, I was instantly back in my body. I was in awful pain and I was crying.

"Oh, my Shrimpito!" said my sister Tencha to me. "Are you okay?"

"I don't know," I said, still crying. "My arm's all twisted and feels broken."

"Good! I hope it is!" yelled Chemo. He looked so upset that I thought he might cry, but he didn't, and instead he just kept on yelling. "You damn fool! I really didn't mean for you to jump when I said jump!"

"Stop that!" shouted Tencha. "I don't know what's going on between you and my little brother, Chemo, but you stop it!"

"Thank God, he's alive!" said Joan.

"Come on, Shrimpito, let's get you in the car so we can go home," said my sister Tencha, helping me to my feet.

"No, I'm walking home," I said, once my sister had me on my feet. "I don't want to get killed again."

"You're not going to get killed," said my sister. "Chemo is going to drive real slow from now on, aren't you, Chemo?"

I turned and looked at my cousin. He was still angry. "No," he said,

"I'm driving like I normally do! I'm not going to let a little kid boss me around!"

"Look," I said, "you guys do what you want. I'll just walk home."

But when I turned, I realized that I didn't know which way to go. My head was throbbing with fire, and I was all confused. As I closed my eyes, it started coming back to me, and I realized that I could go back up the hill towards California Street to go home, or down the hill towards Vista Way and go along the lagoon. Either way, I'd get home. But when I tried to move, I couldn't get my legs to work.

"Look at you! You can't even walk!" said my sister.

"Yes, I can," I said. But my knees wouldn't work going uphill. I guess they'd hit the rocks, too. I turned and started down the hill towards Vista Way, but I was in so much pain, that I wobbled all over the place.

"Shrimpito!" called my sister.

I ignored her. I had to. The whole road looked blurry to me. I could feel tears running down my face, but when I went to wipe them, I found blood on my hand. I began to tremble and feel real cold, but I knew that I couldn't give in to the pain I was feeling.

I continued down the road with tears and blood running down my face, and I was doing pretty well, but then I don't know what happened. One moment, I'd been moving along okay, and the next, I was falling to the ground.

Instantly, my sister Tencha was at my side. "Chemo!" she shouted, picking me up in her arms. "That's it! You're not driving! Joan is, unless you agree to go real slow!"

"All right! All right! We'll do whatever the little king says!"

I was up above myself once again, but this time I wasn't dead. No, I was alive and I could see myself down below as my sister Tencha put me in the backseat of the car and got in with me.

All the way home, Tencha stroked my head. I was happy that I'd chosen to come back and live. This was when I realized that we weren't alone in the car. No, Jesus was also here with us, just as He'd been with me when I'd walked down that long, dark hallway to face the huge Frog. I smiled, thanking Him, and drifted off to sleep.

CHAPTER **fourteen**

Getting home, Chemo immediately started telling my brother Joseph that I was crazy, that I'd jumped out of the car because he wouldn't obey me and slow down.

"I knew what I was doing!" shouted Chemo. He was so upset that he was almost ready to cry. But he didn't. He just kept shouting. "You got to talk to him! He thinks he's a king, and everyone has to obey him! Who the hell does he think he is? And he's willing to die, if he doesn't get his way! He's crazy-*loco* out of his head, *José!*"

Joseph listened very carefully, then he called me inside. Boy, I could feel it. I was in deep trouble. "We need to clean your wounds," my brother said to me once we were inside the house.

My sister Tencha and Joan washed the blood and dirt and gravel off my face. It hurt like hell when they took what was left of my T-shirt off of me. They had to work real carefully to get all the dirt and grime out of my back and shoulders and the back of my head. All this time my brother and Chemo watched. Joseph was in his pajamas and his dark maroon bathrobe.

Finally, Tencha and her girlfriend Joan were done working on me. My brother asked them if he could talk to me alone. We sat down across from one another at the kitchen table. Chemo, Tencha, and Joan walked outside.

"Tell me," said my brother, once we were alone, "did you really mean to jump, or was it an accident?"

"No," I said, "it was no accident. I meant to jump."

"Out of a speeding car?"

"Yes," I said, flashing on Superman, "out of a speeding car."

"*Mundo*, do you realize that you could've been killed?"

"Yes," I said. "I think I was for a little while."

"You were what?"

"I think that I did die," I said.

"You were killed?" he asked.

"Yes, for a little while."

"What makes you think this?"

I shrugged. "I don't know, except, well, it was like I wasn't in my body. I was, instead, up above my body looking down. I think I even saw my own body hit the gravel and go rolling into a ditch."

Hearing this, my brother nodded, and just sat there looking at me, not saying anything for a long time. I said nothing, too, and looked back at my big brother. I'd never realized that Joseph's eyelashes were as long and thick as our father's. They fluttered like butterflies above his large dark eyes. My mother always said that our dad had eyelashes that any woman would die for.

My brother still said nothing. His silence was beginning to make me a little bit nervous. My feet didn't reach the floor when I sat back in my chair. I had to move forward in the chair so I could get my feet on the floor. Still my feet wouldn't stop fidgeting.

"Did you actually use the word 'obey' when you asked Chemo to slow down?" he asked.

I nodded. "Yes," I said. My feet stopped fidgeting. I, too, wondered why I'd used the word "obey."

"I see," said my brother, pulling his chair around the table closer to me. "Do you have any idea what the word 'obey' implies?" he asked.

"Implies?" I said.

"Yes, what the word really means."

I nodded. "Yes, I think so."

"Tell me, what does it mean?"

"It means," I said, "I guess, like when God tells us that we must obey His Ten Commandments, or when people say that we must obey our parents or our teachers."

"Exactly," he said. "Now can you see what it was that this word 'obey' caused?"

I shook my head.

"Think about it. You, a little kid, told Chemo, who's older and bigger than you, and in front of his girlfriend, that he had to 'obey' you. So then, of course, he couldn't do as you said, because he'd look bad."

"But *José*, you don't understand," I said with tears suddenly coming to my eyes. "I wasn't joking! I really, really did see that red truck coming up the hill! I had to get them to slow down, or we would've all been killed!"

Joseph reached out and gently touched me. "I'm not arguing that," he said. "I'm sure you're right and you saw that truck. That's not the issue. *Papa* tells us how he can see the cards that are going to be dealt to him in poker hours before he goes to the game. Duel, his teacher up in Montana, taught our dad how to do this. This is why we know that *papa* will come home a winner from their trip to Las Vegas and the other people won't.

"So you see, the issue isn't if you saw or didn't see that truck coming up the hill. The real issue is, if in the future you are able to somehow see what's going to happen around the next corner again, how can you get people to do as you need them to do, so that a fatal incident can be avoided? Do you see where I'm going?"

My eyes got huge. I'd thought that I was going to get in trouble. Instead, I was being told by my brother how to get even better and stronger at working with these little flashing glimpses of the future, if this ever happened to me again.

"So what word could you have used instead of 'obey'?" asked my brother.

"I don't know," I said. "I'd already said please, and told them about the coming truck, but they still wouldn't take me seriously and slow down."

"You know what you could've said?" said my brother. "You could have said that you needed to pee real bad and asked them to please slow down."

"I did have to pee," I said.

"See, there you have it," continued my brother. "Then I'm sure that Chemo would've immediately slowed down for you if you'd said that you were ready to pee, because he'd have sympathy for you and also because it didn't make him look bad. Does this make sense?"

I nodded. "Yes."

"Good," said my brother, "I'm glad, because you see, *papa* has explained this to me a million times: You never make a man look bad, especially in front of his woman, or you'll always end up in a fight. *Papa* has also told me that many times, when he was a young man, he'd ask the driver to pull over because he had to pee. And once he was out of the car, he then told his friends to go on by themselves, that he had a life to live and he wasn't going to risk it over such a stupid little thing as wild drunk driving. Once, three of his friends got killed. He, too, would've been killed, he told me, if he hadn't gotten himself out of that car. So you see, I'm not saying that your intent was wrong, but how you went about it wasn't very smart. You just can't keep jumping out of speeding cars, even if you think you're Superman," he added, grinning.

I laughed. He'd caught me. Boy, was he smart! This was exactly what I'd begun to think since I'd jumped out of the car, that maybe I was Superman.

"It's good to see you laugh," said my brother. "Life, you know, is fun. But, you do need to learn how to choose your battles. This is what we're taught at the Academy."

"To choose our battles?"

"Yes, when you see something that other people don't see, you have to think ahead and figure out how to get your way, and not waste valuable time."

I nodded.

"Because," he added, "I'm sure that this is going to happen to you again, *Mundo*. It was no accident what happened to you today, just as it was no accident that you were born two years to the day, to the hour that our grandmother Doña Margarita died. You are her, *papa* always says, so the next time you're able to see around the bend in the road of life, you must be prepared to get your way, just as she did in the Revolution."

I nodded again. This I'd heard all my life about my being born two years after our grandmother Doña Margarita died to the day, to the hour. "Then you believe me that I really did see that truck and I had to get them to, well, not obey me, but do as I said?"

"Yes," he said. "You see, I, too, have been seeing a lot of things, now that I've been spending so much time at the hospital." He took a deep breath. "One night I got up and went down the hallway to see this woman who was crying. She was an elderly woman and the doctors didn't know what to do for her. She was dying. I held her hand and stroked her forehead like she was a child. She immediately calmed down and was able to pass over in her sleep so peacefully.

"The night nurse was furious, and the next morning she told the doctors what I'd done. They, too, became upset, telling me I didn't have the authority to visit other patients. That poor nurse and the doctors, they just weren't prepared to accept the simple truth that they aren't in control. No one is in control, *Mundo*. We're all just God's guests for a short time."

I don't know why, but I now asked, "Joseph, are you dying?"

He looked at me straight in the eyes. "Yes, *Mundo*," he said, "I'm dying."

I began to cry. "But, *Chavaboy*, I don't want you dying!"

"Look, you asked, and I told you," he said.

"Yes," I said. "But . . . but when I died, I could see that it was still up to me to stay dead or come back alive."

"Look," said my brother, "I'm glad that you came back, but you see, my situation is very different than yours. Mine has been going on a long time. I've already used up my nine lives. You, on the other hand, have maybe used only one or two."

An icy cold chill went up and down my spine. "Then, it really is my fault that I didn't ride my horse back and tell you not to cross at that place in the swamp, isn't it?" My eyes were overflowing with tears.

"No, *Edmundo,* it's not your fault."

"Yes, it is, you just don't want to tell me!"

"Listen to me," said my brother, "in the last few months of going back and forth to the hospital, I've learned a lot about life, and also about death. And what I've learned is simply this, that everything is already okay, *Mundo,* that *qué será, será, con el favor de Dios.*"

"But, *Chavaboy,* if only I'd rode back, then everything—"

"There are no 'buts' in life, our father always tells us, or then our aunt would have balls and she'd be our uncle."

I laughed. My brother had mixed up the saying. "No, *José,*" I said, "our dad says that for the word 'if,' not 'but.' "

"I think, you're right," he said, also laughing. "I'm glad to see that you've been paying attention, because when it's all said and done, it all comes down to the same thing: There are no 'ifs,' 'shoulds,' 'tries,' 'buts,' or 'maybes' in life. All these are weak words that cause us to remain doubtful and not live to our fullest. At Scripps, over the doctors' protests, I was able to help a lot of people to calm down and gracefully accept their fate."

"So, then, you didn't pay attention to the doctors telling you not to see the other patients?"

"No, I didn't," he said.

I couldn't stop crying. I loved my brother. I just wished that he hadn't told me the truth when I'd asked him if he was dying.

"Look," he said, "who do you think it was that helped you see that red truck coming up the hill? Who do you think helped our two grandmothers through all that starvation and war? Who do you think has been helping the whole of humanity since ever? What do you think *'con el favor de Dios'* really means? It's about asking God to help us with daily miracles."

"Then you think that there really are miracles?"

"Of course, everywhere, everyday. It's just that we close our eyes and don't see these 'red trucks' anymore."

"Well, then," I said, wiping my eyes, "why can't we just ask for a miracle right now and not have you die?"

He looked me in the eyes very quietly, then reached out and took my hand. *"Mundo,* it wasn't your fault," he said. "Can you understand that? It's not anyone's fault, and also, it's not for you or for me or anyone to question." He breathed. "Lying in that hospital, I had a lot of time to think. I began to see little by little that there's a much larger, grander plan going on that we can't see, much less ever understand. Poor *mama* and *papa,* they still think that with their money and modern medicine, they can step in and change the course of my destiny, and sometimes people can, but then, there are other times when we must just relax, let go, and trust *la Mano de Dios."*

Hearing these words, "the Hand of God," an icy cold chill went up and down my spine, and then, strangely enough, once again that soft humming began to vibrate behind my left ear. I didn't understand what my brother was telling me. I really didn't. But the quiet humming got stronger, soothing me like a warm, gentle hand.

"Chavaboy," I said, "do you ever feel a purring right here, behind your left ear?"

"A purring?"

"Yes, like a cat purring or a little massage like when *mama* rubs our forehead when we don't feel well."

"Do you feel this?"

"Sometimes," I said.

"You know, I'd forgotten," he said, smiling. "But now that I think about it, I remember something like that happening to me when I was young. It's a little vibrating behind one of our ears, right? Then it goes across the back of your head to the other ear."

"Yes, exactly!" I said excitedly.

"Then yes, I do remember that happening to me."

"Well, you know what I think that this humming is," I said. "I think that it's the Hand of God massaging us."

He smiled a beautiful smile. "Absolutely. Why not? The more and more that I learn the more and more I realize that God is always with us. It's just that we're too busy to notice." He breathed. "You did nothing wrong," he said once again. "Do you understand, you did nothing wrong. You're just a little kid who was so fascinated by the adventure in that beautiful marsh that you forgot about me, and that's the way it should be. Live, *Mundo,* live, that's what we are all here to do."

Tears were rolling down my face. I had to work hard to not make any crying sounds. I didn't want my brother worrying about me.

"Do you understand?" he asked again.

I couldn't talk, but finally I was able to say, "Yes, I think so."

"Good," he said, "because what I've come to understand lately is that it's not control or money or new inventions that are needed in this world. What is really needed is so simple. It's patience," he said, taking a deep breath. "Patience, compassion, love, forgiveness, and understanding, so don't you dare blame yourself or anyone else for anything," he added.

"But *Chavaboy,*" I said, the stream of tears still running down my face, "I don't want you dying! *Papa* told me that no one is leaving our *familia!*" I shouted.

"And he's right, no one is leaving," he said. "Remember how you saw your own body rolling across the road into the ditch?"

"Yes," I said.

"Well, like that, I'll always be watching over you, *hermanito.* I'm not leaving. I'll always be here, just above you, watching out for you."

"Really? You promise."

"*Te juro,*" said my brother.

I drew close and hugged my big brother Joseph with all my heart and soul. Then I'll never forget. After we were done hugging, he sat back and turned on the record player at his side. The song "Ghost Riders in the Sky" began to play. For over a month now, this was the song that my brother kept playing, over and over again.

An old cowpoke went ridin' out one dark and windy day,
Upon a ridge he rested as he went along his way
When all at once a mighty herd of red-eyed cows he saw
A' plowin' through the ragged skies, and up a cloudy draw.

Yippee-aye-aaa, Yippee aye-ooh
Ghost Riders in the Sky

Their brands were still on fire and their hooves were made of steel
Their horns were black and shiny and their hot breath he could feel
A bolt of fear went through him as they thundered through the sky
He saw the riders comin' hard . . . and he heard their mournful cry

Yippee aye-aaa, Yippee aye-ooh
Ghost Riders in the Sky

Their faces gaunt their eyes were blurred their shirts all soaked
    with sweat
They're ridin' hard to catch that herd but they aint caught 'em yet
'Cause they've got to ride forever in the range up in the sky
On horses snorting fire as they ride on hear their awful cry

Yippee aye-aaa, Yippee aye-ooh
Ghost Riders in the Sky.

The riders loped on by him he heard one call his name
If you want to save your soul from hell a-riding on our range
Then cowboy change your ways today or with us you will ride
A-tryin' to catch this devil herd across these endless skies.

Yippee aye-aaa, Yippee aye-ooh,
Ghost Riders in the Sky.
Ghost Riders in the Sky.
Ghost Riders in the Sky.

CHAPTER **fifteen**

The following day our parents returned from Las Vegas. My dad and mom wanted to know what had happened to me. My face and shoulder were all cut up. I didn't know what to say. My brother told them that I'd just had a little accident and everything was okay.

Our dad brought a big fat, heavy canvas bag into the house, and turned it upside down. Silver dollars spilled out all over the place! He'd won, and won big, and everyone else had lost!

Archie and George helped themselves to the liquor in the bar. Everyone seemed to be having a great time, except *tía* Tota. She seemed upset about something. George and Vera went home to be with their kids. My father suggested that the rest of us go out for Chinese dinner.

"Oh no, you don't, Salvador!" shouted our *tía* Tota. "You're just trying to trick me and poison me with fish again!"

"Hell, I wasn't even thinking about you, one way or the other. I was thinking of Lupe. I don't want Lupe making dinner for all of us."

"Thinking of Lupe!" yelled our *tía*. "When you do you ever think of her! All you men ever think of is your liquor and gambling, and leave us women alone in our rooms!"

"We took you to the shows," said our dad.

"Yes, to that show with naked-breasted women where you men all went crazy!"

"But you and Lupe chose that show!" yelled our dad.

And so it was another fight between my father and our aunt. Some things just never changed.

. . .

Two days later, they took my brother Joseph to the hospital while I was at school. When I got home and found out, I was pissed! My brother had tried so hard to not give in to his sickness, but in the end, God had forsaken him!

My dad kept saying it was all Dr. Hoskins's fault. If the damn fool hadn't been drinking all the time, he would've spotted Joseph's condition months ago, and then none of this would have happened. But I knew that it wasn't just Dr. Hoskins who'd failed my brother. I, too, had failed him when I hadn't ridden back to help him cross the marshy inlet.

I was mad, and at school, I began getting mean. Instantly, I began winning at marbles once again. I'd wipe out whole pots, keeping all the marbles. Then a new girl came to school whose father was a minister. She was real pretty. Her name was Judy and she asked me and a good friend of mine named Dennis if we had any money. We both said that we did.

"How'd you get your money?" she asked. "My parents won't give me any, saying that money is the root of all evil."

"I get an allowance of fifty cents a week," said Dennis.

"I blackmail people," I said.

"Really? You blackmail people!" she said to me. "How do you do that?"

"Well, first you need to find out about something that someone is trying to hide, or that they want to do without permission. Then you charge them to not tell on them."

Her eyes got big. She loved it.

"Would you ever blackmail me?" asked Judy.

"No, you got no money," I said. "That's another thing that I've found out. It's not worth your time to blackmail people who have no money."

"That makes sense," she said. Then she proposed something that I'd never thought of as a way of making money. "Look," she said, "if you and Dennis meet me after school, I'll pull up my dress and show you my underwear for five cents each."

"Okay," said Dennis, anxiously.

"Just wait," I said, "what if Dennis and I pull down our pants and show you our underwear, then charge you five cents each."

She laughed. "But I don't care about seeing your underwear," she said. "I see my brother's all the time, and it's just plain white. But my underwear," she added with a big grin, "is pink and has little flowers and you can see a lot of my legs and my belly button, too."

I didn't like this one little bit! I thought it was totally unfair that we couldn't get money for pulling down our pants. But I had to admit that getting to see her underwear did sound a lot more exciting than us showing her ours.

Finally, Dennis and I agreed, and we met Judy down at the big eucalyptus trees behind the market that old man Hightower had just built. First, she took our nickels, before showing us anything, but then when she pulled up her dress, she'd only let us see the front part of her. I wanted to see the back of her underwear, too, but she said that to see the back of her underwear would cost us another nickel. I told her that I hadn't brought any more nickels. She told me that I could bring my money tomorrow.

That night, our mother didn't come home from the hospital with our father. Our dad told my little sister Linda and me that our mother wouldn't be coming home to eat dinner with us anymore. She would now be staying down in La Jolla day and night, either at the hospital or at a hotel where she'd rented a room down the street from Scripps. I began to cry. This was awful. My brother Joseph was really dying.

The next day at school, Judy immediately wanted to know if I'd brought my other nickel so she could show me the back of her underwear after school. But I didn't want to talk about money or underwear, so I told her to talk to her dad, that he was a minister so he had to have money. She told me that her father didn't have money, that they were poor. Then Judy got a strange look in her eyes and began to stroke Dennis's cheek. She told Dennis that maybe she wouldn't even charge him anything to see her after school, if he became her boyfriend. Seeing this, I quickly spoke up. After all, Dennis was my friend.

"Look," I said, "whatever you do, don't ever let her get on top, because if you do, then you'll get all embarrassed and angry if people find out that you were on the bottom."

"The bottom of what?" asked Dennis.

"The bottom of her," I said, "and her on top of you like a rooster does to a chicken, going up and down."

"Really," said the minister's daughter, "guys get embarrassed and angry if the girl's on top?"

"Yeah," I said, feeling proud that I knew so much. "I saw it happen once with my—I can't tell you with who, but it's true."

"Did you blackmail them?" asked Judy.

"Yes."

"Really. How much did you get?"

"A dollar," I said.

"A whole dollar!" she yelped. "Wow! How can I get into this blackmail, too?"

"I don't know," I said.

"Look, you show my boyfriend Dennis and me how to do blackmail," she said, "and I'll even let you touch me."

"Where?"

"You know."

"No, I don't."

They both started laughing so much that I got mad and didn't meet with them after school. I wanted to get home to find out how my brother Joseph was doing. I wanted to ask my dad if I could go down to Scripps with him and see my brother. But that night, when our dad got home, my little sister and I could both see that he was in no mood to talk. He smelled of whiskey. We ate in silence at the dinner table that night while *Rosa* served us. I could just feel it; things were really getting bad for my brother Joseph. That night I prayed for God to please spare my brother's life, just as I'd heard my mother pray so many times in the last year.

The next day at school, Judy and Dennis met with me at recess. All they wanted to know was how they could get into my blackmail business. But no matter how much I explained to them that it was better for them to start their own—because I was blackmailing my own family—they wouldn't take no for an answer.

"Look," I finally said, trying to get rid of them, "just snoop around here on the teachers and find them doing something they shouldn't be doing, so you can blackmail them."

A few days later, Dennis and Judy came to me all excited and said they'd kept their eyes open, and yesterday they'd followed the fourth-grade teacher, the youngest teacher at our whole school, and they'd seen her meet someone by that little cemetery just south of school over looking the duck lagoon between Oceanside and Carlsbad.

"They were holding hands," said Judy all excited, "and walking around the graves. They kissed next to one grave!" she added.

"So how much do you think we should ask her for?" asked Dennis.

"Were they glancing around like they were afraid of somebody seeing them?" I asked.

"No, not that I could see," said Dennis.

"Look," I said, "I don't think you have enough."

"But they were really kissing, and next to a grave!" said Judy. "So that has to be worth something."

"Not if they're married," I said, "or boyfriend and girlfriend. Just holding hands and kissing isn't going to get you any money. You see, for blackmail to really work, the people got to be real scared of somebody finding out what they've done. That's how the priest works it at church. He gets everybody feeling real bad about what they've done all week, then he scares the sssh—I mean, the crap out of them with hell and damnation, and this is when people pull their money out of their pockets and put it in the basket. Ask your dad," I said to Judy, "I bet he's really good at blackmailing people into giving him money on Sundays."

"No, he isn't," she said, "like I told you, we're poor."

"But that doesn't make sense," I said. "Maybe you should spy on your dad. Because our priest at Saint Mary's in Oceanside, he's old, but he sure knows how to scare the ssh—I mean, the crap out of people, and get them to give money on Sunday and even pay for masses, too."

"You don't have to say 'crap' because of me," Judy said. "You can say 'shit,' if shit is what you really want to say."

"Really?"

"Sure. Why not? I say all the bad words I can when my dad isn't around. And my mom says some, too, when he isn't listening to her."

"I'll be damned—I mean, blessed. No, I mean damn, DAMN, DAMN," I said. And why shouldn't I say "damn" instead of "blessed?" I said to myself.

Chavaboy was getting worse every day.

At home our silent dinners continued and our dad just seemed to be getting worse and worse with the smell of whiskey. Our mother hadn't been home in a week, and Linda and I missed her so much.

I began lighting the candles at my mother's little altar just outside my parents' bedroom and I had Linda kneel down with me so that we could pray together for our brother and mother, too. The candles would flicker in the hallway darkness and I swear that I'd sometimes see something move out of the corner of my eye, but it didn't scare me. I knew who it was. We weren't alone. Jesus was here with my sister and me, praying, too.

And at school, no matter what I'd say, I couldn't get Dennis and Judy to talk about anything but money and blackmail. Still, they just couldn't seem to get enough on anyone so they could get them to pay. Yet they kept trying, and who could blame them, especially Judy, who was always so broke that she couldn't even buy herself a piece of candy if she wanted one.

Once Judy and Dennis and I went into Hightower's Market and Judy told me to go up to old man Hightower and ask him if he had Prince Albert in a can, while she and Dennis checked some things out in the back of the store. I went up to Mr. Hightower and asked if he had Prince Albert in a can. He said sure, reaching behind himself and getting this greenish-looking can of tobacco.

"But you have to get your father to come to buy it," he said.

Just then, Judy and Dennis came running by me, yelling for Mr. Hightower to let poor Prince Albert out of the can, and they ran out the front door.

Mr. Hightower got mad.

I couldn't figure out what was going on, so I took off, too. Getting on my bike, I was able to catch up with Judy and Dennis down by Buccaneer Beach. They were laughing and laughing and they had all these candy bars and little sacks of peanuts and Fritos. Suddenly, I realized that they'd sent me to talk to Mr. Hightower so they could steal these things from the back of the store. I was shocked. But they didn't seem upset and offered me a candy bar, saying that it had been great freeing Prince Albert from captivity.

When I told them that stealing was wrong, they just laughed and asked me, what did I think blackmail was, God's work? Hearing this, I got an icy-bad feeling. Maybe God was punishing my brother for all the horrible things that I was doing in my life. My mother was right. God was taking Joseph away from us because we were such a bad, horrible, no-good *familia!*

I began to cry, but didn't want Judy and Dennis to see me crying, so I got on my bike and pedaled home as fast as I could. That night, our dad didn't come home in time for dinner, so Linda and I ate dinner with *Rosa* and *Emilio* and their little boy, *Carlitos*. It was good to hear people talk and laugh at the dinner table again. Our dad did get home in time to put us to bed. But he didn't know how to sing us to sleep or rub our foreheads real good like our mother. Also, his eyes looked all red and he smelled terrible.

The next day at school, Dennis and I were playing marbles with the guys when Judy came up. She was all upset. I could see it in her eyes, but she wouldn't tell us what had happened. All I knew was that lately, she'd begun to spy on her own father, so she could get him to give her an allowance. But he wouldn't because he kept telling her that money was the root of all evil. Hell, I'd been told the opposite by my dad. He'd told me that being penniless was the root of all evil, and any fool who didn't know this just hadn't lived through starvation.

That same day after school, I saw Judy looking so brokenhearted that I gave her a dollar. "And you don't have to show me nothing," I said to her. "I'm giving you this money because I like you, and you're my friend, Judy."

Hearing this, she took the dollar, looked at me in the eyes, then gave me a great big hug and a kiss on the cheek, too, but not the way she kissed her boyfriend Dennis. Boy, it felt good! It was the first kiss that I'd ever gotten from a girl who wasn't a relative.

A few days later, I could smell something really wrong the moment I walked out of the classroom. The school grounds were empty. Once again, I was being kept after school every day, so my teacher could try and teach me how to read before the end of the year.

I was walking quickly across the empty asphalt to get my Schwinn, so I could go home. This was when I saw three guys come out from behind a small pepper tree. My heart started pounding. Something was really wrong, I could smell it. These guys weren't my friends. But of course, I knew them all. Two were in my present third-grade class and one was in the fourth grade, in the same classroom that I'd been in before I'd flunked.

They circled around me. They had funny little grins on their faces, but I couldn't figure out what it was that they wanted. Then it hit me, they were probably mad at me for winning at marbles so much lately. But before I could say anything, they all rushed at me, and began hitting me as hard as they could.

I was so shocked that I didn't hit back at first, but then one boy said something about me being a Mexican, and hearing this, I got real mad, and started hitting back with all my might, and kicking, too!

And when they knocked me to the ground and I saw I didn't have a chance, I started grabbing and pulling them to me and biting them as hard as I could, just like a wild dog. This was when the screaming started and they got off of me as fast as they could.

I got back on my feet, too. "I WON THOSE MARBLES FAIR AND SQUARE!" I yelled. "You're just sore losers!"

"You bit me!" said one boy, crying. "I'm bleeding, you stupid Mexican! Don't you know teeth carry disease!"

"What was I supposed to do! You guys started it!"

"No, we didn't!" said another. "You did it! It's all your fault! Next year more Mexicans are coming to our school because of you! Why don't you people all stay in Pas-ol-eee Town where you belong!"

I was all confused. I'd thought that this was about me winning at marbles. "You mean you guys want to beat me up because other Mexicans are coming to this school next year?"

"YOU'RE DAMN RIGHT!" yelled one kid. He was new. "I'm from Texas! And my pa told me that we got to put you all in your place! That he don't care how big your father's *hacienda* is, you're still a damn, worthless, *chile* belly greaser!" Then he said, just like Gus had, "Remember the Alamo!"

All the blood left my face. Suddenly, all these flashes of Ramón and us kids getting beat up by the teachers and other kids at that other school across town came rushing to my mind. Was God also punishing me for being Mexican? I began to cry. Maybe this was really what it was all about? Maybe this was what my mother had meant when she'd told my dad that God was punishing Joseph for their sins, the sins of us all being *Mexicanos*. If this was true, then my *tía Tota* was right to say that she was glad she'd never had any kids. After all, we Mexicans were no good people, so it was wrong for us to have kids. I got on my bike and went home crying.

But I can tell you, all that didn't shock me as much as the next day when I came out of the classroom to get on my Schwinn and there was Dennis, my best friend, and he, too, was with these other three guys waiting to beat me up.

Well, this time, when they started circling me, I didn't wait for them to jump me. Feeling SO-O-O BETRAYED seeing Dennis with them, I attacked the way my dad had told me he'd attacked a desert tiger that had been getting ready to leap on him when he'd been a little kid back in old Mexico.

I raised up my arms and SCREAMED at the top of my lungs, startling them, hitting the biggest guy who was a fourth grader in the nose as hard as I could. Then I grabbed the next biggest kid from Texas, and yanked him to me, and bit him in the face. These two were now screaming as I was hitting the other two. They now all took off running, and I screamed and screamed at them. "Come back! COME BACK! I'LL FIGHT! I'LL FIGHT, *cabrones!* REMEMBER THE ALAMO!"

Hell, I didn't even know what the Alamo was, but it had felt good to yell this. But I never really hit Dennis too hard, even though he took a few swings at me, because I could see that he wasn't really into hurting me. In fact, it was easy for me to see that he'd felt ashamed.

The following day, it got even worse. Once more the school grounds were empty by the time I came out of my extra half hour of reading, and I was a sitting duck. This time there were five guys waiting for me and their smell of wanting a fight was really strong. Two of them were great big guys whom I didn't know, so I figured that they were maybe even fifth graders.

They hurt me real bad this afternoon. They kept kicking me once they got me down. In the end, I'd hit and scratched and bit so much that all five boys backed away from me, not wanting to fight me anymore.

I'd held my own. I hadn't backed down. But you wouldn't know it by the way I looked and felt. I couldn't even raise my arms high enough to get on my bike to ride home, I was in so much pain. And my head was all dizzy from the kicking. I then remembered, like in a foggy far away dream, that my brother had gotten his sickness from a football injury. I wondered if I, too, was now going to end up at Scripps.

I cried all the way home once again. It wasn't my fault that I'd been born a Mexican and that other Mexicans were coming to school next year.

"God," I said, "why did you even create any Mexicans or Blacks, if we're all such bad, no good people? Eh, You tell me that."

I had to stop and rest several times before I could make it home, I was so beaten up. I felt much better once I got to the gates of our *rancho grande*. I was home. This was our place, our world, our reservation. And here everyone spoke Spanish. So maybe here it was okay to be *un Mexicano*.

That night our dad didn't come back from La Jolla. He called to say that he was staying overnight with our mother to keep her and our brother company. *Rosa* and *Emilio* attended to my wounds, wrapped my ribs, and lit candles and prayed for the Virgin Mary to protect me when I told them what had happened.

Emilio then told me how he'd also been beaten by the cops in Texas when he'd complained that his boss wouldn't pay him. I said an extra prayer, asking for Ramón and the other guys from *Pozole* Town to please come and help me at this school in South Oceanside because I just couldn't keep doing it alone another day. Tomorrow they'd be sure to kill me.

But when I got to school the next day, the strangest thing happened. I was immediately called into the principal's office and told that several parents had come to school complaining about my hitting their children.

"Me? Hitting their kids!" I said in astonishment. And I tried to explain to the principal that they'd attacked me, but he got real angry and just cut me off, saying that I had to stop hitting and biting people at his school.

"Maybe this is normal behavior where you came from, south of the border, but here it isn't!" he said, yelling at me.

"But they're the ones who picked on me!" said I, beginning to cry.

"Listen here!" said the principal, getting out of his chair and coming around his desk. I crouched down, expecting him to hit me, but he didn't. He just shook his finger at me and said, "Four boys can't all be wrong, and you right! You ever hit anyone again, and I'll personally take a ruler to you! Get it?"

The tears were pouring from my eyes. This felt so totally unfair. But the principal just couldn't seem to get it through his head what I was trying to say. Then I couldn't believe it. He told me the very same thing that the kids had told me.

"Next year more of you Mexicans will be coming to this school, and your kind of people need to know your place!"

That morning, I was crying when I walked out of the principal's office. And who could I talk to? No one. My parents were down in La Jolla with my brother and they had no time for me.

Then I saw them, the guys who'd beat me up, including Dennis. They were across the asphalt waiting for me. But when they saw me crying, they started laughing, and making catcalls. At least they didn't beat me up that afternoon.

The following day, Dennis and Judy came up to me. I was sitting by myself. School wasn't the same anymore. It had all changed in just three days. Now I kept away from everyone, or more precisely, everyone kept away from me. I sat by myself, eating lunch. After all, I was the only Mexican at school who looked Mexican. The other kid who I knew was a *Mexicano* was light-skinned and kept telling everyone that he was Spanish just like those other kids had done across town.

"Dennis has something to say to you," said Judy to me.

I could see that Dennis really didn't want to say it.

"Dennis, you do it!" said Judy. "Or I won't be your girlfriend anymore!"

"I'm sorry that I hit you," said Dennis to me. "It was wrong for us to gang up on you. It's really not your fault you're Mexican or that other Mexicans are coming to school next year."

"That's not what I told you to say!" snapped Judy. "I told you to tell him that he's a good, decent person and you were all cowards to gang up on him, and you're sorry and you'll never do it again!"

He glanced at her, saw her anger, then turned back to me. "Judy's right," he said. "We were cowards to gang up on you and you're a good, decent guy, and I'll never do it again."

Judy now came forward and hugged me. "I talked to all of them," she said. "They're never going to pick on you again."

"They're not?"

"No, I blackmailed them," she said with excitement.

"How'd you do that?" I asked.

"I told them that I'd tell our teacher that they all pulled my underwear off of me if they didn't do as I say."

"Did they really do that?" I asked, my heart suddenly exploding, because Judy was my friend, and so I'd really fight them all now!

"Do what?"

"Pull down your underwear?"

"Oh, no," she said, "they wouldn't dare! They're cowards. I only said that I'd say that to get them to obey me."

I laughed. Judy had just used the word "obey," too! And, also, she'd taken blackmail to an all-new level. I'd never thought of just making up stuff, when you couldn't find what you needed so you could blackmail people. She was really smart. Her dad could learn a lot from her about passing the basket on Sundays.

A few days later, another big surprise came to me. This kid named Gary, whose father was an ex-Marine and was now a local fireman, came and asked me if he could fight with me. I'd never had some one come up and ask for permission to fight me. I didn't know what to say.

"Look," he said, "my dad explained to me that a sneak attack like those guys did to you is only for cowards. So he told me that since I'm a new guy at school and you're the strongest and toughest kid in school, I should see if I can beat you, but you know, in a fair fight, then everyone will respect me. But no biting or pulling hair. Just wrestling and boxing."

"You mean, your dad told you that you should come to school and fight me?"

He nodded. "Yeah, sure."

This was all so different from everything I'd ever been told. My dad had told me that only fools fought, because there were no odds or money in fighting, and that real men only did it as a last resort.

"You see," he continued, "it's our duty, as Americans, to put your people in your place, but not by a sneak attack like at Pearl Harbor."

Suddenly, I knew that this guy Gary was being truthful with me, because I knew about Pearl Harbor, and I also knew that most of these kids' fathers were ex-Marine or ex-Navy men who'd been involved in the war, and so they probably all thought the same way.

I suddenly felt so bad for all these kids, and for me, too. Maybe my mother was absolutely right, and we not only got our parents' sins passed down to us, but also their way of thinking.

"Okay," I said, "I'll fight you, but it's got to be a good, fair fight. Nothing dirty. And just the two of us and no one watching." I didn't want to take a chance of some other guys watching and then jumping in to help him when I started winning.

He put out his hand to shake on it. I took his hand, and we shook on it. That same day, I met Gary by the tall eucalyptus trees at the far southeast corner of the school grounds. School was over. Gary had waited for me to do my half hour of reading after school.

Immediately, Gary took off his shirt, I guess to show me how muscular he was. But I thought that this was a dumb thing to do, because once we got down on the ground, rolling in the dirt and rock and leaves, it was going to hurt his back. But he didn't seem to care and he came at me with his fists up in front of him like he knew how to box. I didn't, so I didn't even try. I just kept my hands open, palms out, and took his first few punches in my open hands.

Then I grabbed his next hit, jerked him to me—which surprised him—and the moment that we tied up, I could feel that he was way stronger than the other guys that I'd fought, but I could also tell that, compared to me, he had no real strength.

I guess that all those afternoons of moving hay and feeding livestock had turned me into a very strong kid. Quickly, easily, I took him down with a headlock and I could've finished him off just like that, making him cry, but he was trying so hard, grunting and pushing, trying his best to get free from me, that I finally let him go. Instantly, he came alive!

I let him get on top of me and beat me up, and why I did this, I had no idea. But then, within a couple of days, I realized that I'd been a real genius to listen to the Voice inside of me that told me to let him beat me. Now he was telling everyone that he was the champ, the strongest, toughest kid in the whole school all the way up to the fifth grade. And when people doubted his word, that he'd beat me in a fair fight, I'd backed him up, saying that he really had, that he was the champ, and everyone was now so happy to have him as their champion that I was forgotten.

It was beautiful. I was free. And I guessed, that I was freed, because they figured that I'd in been put in my place, and so now everything was as it should be. But what my place was, I still didn't know, so I just kept to myself, watching and learning. I even stopped playing marbles.

CHAPTER **sixteen**

Howling! HOWLING! My brother Joseph's dog Shep was going crazy-*loco*, racing around and around the house, howling to the Heavens! It was late at night and I was sound asleep. Both our father and mother had now been down in La Jolla with our brother for days on end.

I could hear Shep racing around and around the house, howling like crazy. Finally, I got up to go see what was the matter. After all, Shep was a very smart dog and didn't bark just for the hell of it. I pulled on my cowboy boots, and reached for my big, new long bow, but then settled on my trusty, old Red Rider BB gun instead, and ran out the front door.

The Mother Moon was full, and I knew that when the moon was full like this, wild animals often came down from the hills and went to the sea through the canyon below our home. But Shep wasn't barking at the canyon behind our home. No, he was howling as he raced around our *casa grande,* and I couldn't see anything going on near our home. As far as I could tell, Shep was just chasing after his own shadow.

"Shep!" I shouted. "It's all right! Calm down! Nothing's going on, boy!"

I wanted him to calm down so I could pet him and he'd feel better, but he wouldn't come near me. He just kept racing around and around, howling like he was crazy-*loco.* Shep was always so smart, so level-headed, and usually listened to me, so I couldn't figure out what was going on. He was acting really strange. No matter how much I called to him, he just ignored me and kept racing around and around the house, howling something fierce.

My little sister Linda finally woke up and came outside to see what was the matter, too. Then *Rosa* came outside to check on us.

"Is everything all right?" asked *Rosa* in Spanish. Linda and I spoke only Spanish with *Rosa*.

"Yes," I said, "it's just that my brother's dog has gone crazy. He's barking at nothing, and won't come to me."

"He's not barking at nothing," *Rosa* said to me. "He's barking, because"—tears came to her eyes—"your brother is dying."

"You mean, like at this very moment?"

"Yes," she said.

I was shocked. I didn't know what to think. "But how can Shep know that my brother is dying right now?" I asked. "My brother is all the way down in La Jolla, more than thirty miles away. You got to be WRONG! YOU JUST DON'T KNOW WHAT YOU'RE TALKING ABOUT!" I shouted at her.

Rosa's husband *Emilio* suddenly came out of the darkness, and I could now see that several of our other ranch hands were all sitting quietly under the huge pepper tree, poking at the dirt with sticks as they sat on the ground.

"*Rosa* is right," said *Emilio* to me. "It's time that you know. Your brother is dying. That's why his dog is going crazy. He's always loved your brother very much, so his heart is breaking."

"But *Emilio*," I said, "only yesterday, I talked to our dad on the phone and he told me that our brother is getting better, that he's eating again for the first time in weeks," I added.

My heart was pounding a million miles an hour. Seeing how upset I was, *Emilio* glanced at his wife *Rosa*, then he came close to me and squatted down so we were at eye level. He put his hand on my shoulder.

"Listen closely," he said, "animals don't know how to lie, or how to pretend. And the soul, she knows no distance, and love speaks through the heart, and this dog is telling us of his heart breaking for your brother, because he is dying."

"OH, NO!" I yelled. "PLEASE! MY BROTHER ISN'T DYING! Please, Shep has got to stop howling, and saying this! SHEP!" I screamed. "You stop telling people that my brother Joseph is dying! You hear me, you stop saying that or . . . or . . . or I'll shoot you!"

But no matter how much I yelled at Shep or threatened him, he wouldn't stop. I began to cry, too. Now my heart was also breaking, just like Shep's.

The workmen never said a word. They just sat there under the huge pepper tree, poking at the dirt with sticks. Finally, *Rosa* took my sister and me back inside.

All that night, Shep kept howling and howling in a ghostly, eerie howl, and twice, I swear, the Mother Moon came down from the Heavens, and she poked her face right up close to my window.

Then, in the early morning hours of the night, Shep suddenly stopped howling, just like that, and when I went outside to see what had happened, I was told by *Rosa* and *Emilio* that my brother had just died at this exact moment, and so Shep had taken off for the hills to intercept his Soul.

I started crying like I'd never cried before. And I didn't go back to bed. I was feeling too scared and all confused. The whole night had been like a crazy, wild dream.

Later that same morning, our parents came home. Our mother hadn't been home to see us for nearly two weeks. All this time she'd spent with our brother down in La Jolla. My sister and I rushed out, being so happy to see our *mama*. But when I saw how swollen her face was from all her crying, I stopped dead in my tracks. My little sister Linda didn't. She ran up to her mother with her arms open. Our mother never saw her. She just walked right past us and went into the house without saying a word. Our father, he was the one who stayed outside with my sister and me. He told us that our brother Joseph had died.

I SCREAMED at the top of my lungs! Our dad took my sister and me in his arms. My sister Linda and I cried and cried. And years later, my sister would tell me that no, she hadn't really cried. That she'd only pretended to cry because she still hadn't really known what people were supposed to do when they were told that someone of the *familia* had died.

All morning, my sister and I stayed outside by the apricot tree, being quiet and real sad. Our dad went inside the house to be with our mother. Later that same day, I told *Emilio* and *Rosa* that they were right, that my brother had died as they'd said. But they told me that they already knew.

"When we saw Shep take off this morning for the far hills to intercept your brother's soul," said *Rosa,* "we looked up and sure enough, a little while later, we saw a shooting star going across the heavens, so we knew that like all good dogs, Shep had caught his master's soul, guiding him back up to God," she said, making the sign of the cross over herself.

Hearing these words, I got a powerful image of Shep, my brother's dog, running up to the highest hilltop that he could find and leaping into the Father Sky and becoming that Shooting Star, as he joined in with my brother's Sacred Soul, so that together they could make their way back to Heaven.

Late that afternoon, I saddled up to go and look for Shep's body. *Emilio* sent two of our cowhands with me, and all afternoon we kept looking, but we never found his body. One of our *vaqueros* told me that animals didn't necessarily leave their bodies behind like humans, that often animals took their Earth Bodies with them when they passed over to the Spirit World, just as they'd taught Our Lord Jesus how to do.

When I heard this, a great peace came over me, and the three of us rode up and down the hills and down into the deep canyons until it was dark. We were on that tall *mesa* just south of the railroad tracks and west of the cemetery on El Camino when the first star of the night came out. We hadn't been able to locate Shep's body no matter how hard we searched. The Mother Moon was out, too, and she was looking so close and all alive. My eyes started watering, but I wasn't crying inside. No, inside I was happy, because I just knew that this was the exact spot where Shep, the smartest human being that I'd ever met, had leaped into the Father Sky to intercept my brother Joseph's Sacred Soul.

I rubbed the tears out of my eyes and that little, quiet humming began behind my left ear. Then I saw it, I saw it so clearly, that this was now a Sacred Place, too, just like that Golden Eye-Opening in Heaven, or that pew in church where I'd sat listening to the Silence of God all around me.

I took a deep breath and glanced around at the dark land and the sky lighting up with stars and the Mother Moon. I came to realize that the *mesa* we were on had to probably be the highest point in all of South Oceanside. My eyes continued watering—I was so sad, and yet also happy.

"Thank you, Shep," I said. "*Muchas gracias* for guiding my brother *Chavaboy* back up to Heaven. Also, thank you very much for taking the time to befriend me while you were here on Earth. I'll never forget you, Shep, I'll never forget you. And thank you, too, Joseph," I said, making the sign of the cross over myself, "for being the greatest brother anyone could ever have. I love you. I love you both with all my heart and soul!"

Having said this, I just knew that my brother had kept his word and he was here, right now, just above me, watching over me. All the way home that night, I could feel Joseph and Shep staying real close to me, guiding my every step in the darkness.

I wasn't traveling alone. I'd never be traveling alone again no matter where I went for the rest of my life, because my big brother Joseph had kept his word and he would now always be here with me, just above me, as I'd been above myself when I'd jumped out of that speeding car.

Getting home, I immediately tried to explain to my parents what had happened on the tall *mesa* east of us, so that they could feel better, but they wouldn't hear me.

But when I told *Emilio* and *Rosa* what had happened—at the stables before going home—they quickly understood and told me that Shep's body had probably gone up in smoke when his Soul had gone over to the Other Side to be with my brother's Soul, because this was what Souls did, working together in Harmony just like all the rest of God's Creation.

"Animals, you see," *Rosa* explained to me, "they can still do this at will, much easier than humans, because they still haven't learned how to question, and so *amor de la alma* is still their basis for living, just like as it was for Lord God Jesus," she added, making the sign of the cross

over herself, then kissing the back of her thumb, which was folded on
top of her bent index finger.

The day of my brother's funeral, Colonel Atkinson brought a
busload of uniformed cadets from the Army Navy Academy in Carlsbad
to march at the funeral and blow the bugle. Well over five hundred
people attended my brother's funeral. I got to wear the little gray-and-
maroon suit that my godparents Manuelita and Vincente had given
me for my first communion. One cadet, a friend of my brother's, tied
my tie for me. They buried my brother at the north end of the Oceanside cemetery on El Camino, just a little way away from a great big
white cross with Jesus. His mother Mary was praying at the foot of the
cross.

Suddenly, all the things that had made no sense to me were very
clear. I mean, here was Jesus hanging on the cross with His Holy
Mother praying for him, and here was my own mother just a few feet
away, praying and crying as they lowered her son's body into the
ground.

Tears of joy came to my eyes and I turned and looked across the little
valley to the west of us. I'd never realized that the big tall *mesa*, where I
figured that Shep had leaped into the Father Sky to intercept my
brother's Soul, was just across this valley from the cemetery. I smiled.
This was so beautiful. My brother Joseph would now be here forever
alongside Jesus and Mary, and anytime he wished, he could look from
his grave site to the west and see where it was that his dog had inter-
cepted his Soul to guide him back to Heaven. It was easy for me to now
understand what my brother had been talking about when he'd said
that there was a much larger, grander plan going on than we know
nothing about. Jesus and Mary, my mother and Joseph—here they were
all together, like *familia*.

I got so excited that I wanted to tell my mother about this, but she
wouldn't stop crying long enough for me to tell her.

Then the burial was over and we began walking back to our cars, when our *tía Tota* once more told everyone why she'd never wanted children, right in front of Linda and Tencha and me.

"Having children is terrible!" shouted our aunt, loud enough for everyone to hear. "It's so painful to bring them into the world, and then it's so much pain to watch them die before you die! Lupe and I both saw this happen to us in the Revolution over and over again with our beloved mother, so I told Lupe, don't have children," said our aunt, crying and crying. "But no, she just wouldn't listen to me, and now look at her suffering so much!"

"Carlota!" said our father, glancing at us kids. "Could you please just shut your mouth for once in your *PINCHE* LIFE!"

"IT'S YOU MEN WHO SHOULD SHUT YOUR PANTS," yelled our aunt, "always forcing us to have more and more children when all we want is a little love and kindness!"

"KINDNESS!" bellowed our dad, grabbing our *tía Tota* by the throat, "you've never had a *pinche* word of kindness for anyone but yourself—you LOUD-MOUTH STUPID, SELFISH WOMAN! Don't you see how your words cut Lupe's heart and hurt our children?"

"Salvador," said our mother, pulling our dad off our aunt, "you leave Carlota alone! Even she, in her own way, means well! Please, no more fighting between you two today!"

Getting home, we found out my mother didn't have enough silver or crystal to serve everyone, so our *tía Tota* brought her silver and crystal over from her home to help feed everyone who'd come over after the funeral. I'll never forget that night, when I was helping to clean the kitchen with her and my cousins *Isabela* and *Loti*. My *tía Tota* took me aside.

"Look," she said to me, holding a fork from her set next to my mother's, "my silverware is bigger, better, and fancier, and more expensive than your mother's. Just because you live in a great home, don't you think that the rest of us don't have fine things. Archie could build me a house bigger than this, and on the beach, too, but he won't do it, because he's not a show-off like your abusive father. Lupe should never have had kids!" she added angrily.

I was stunned. I felt like taking her bigger, more expensive fork and driving it into her belly where the flesh was soft. But I didn't. I just stood there, listening to her showing me why her silverware was larger and better than my mother's. After all, maybe she was right and none of us Mexican kids should have ever been born.

Then, on the seventh day after my brother had passed over, *Emilio* told me that Joseph and Shep had made it home. "This morning, when I got up, I looked up at the sky and I saw that the morning star was bigger and brighter than I've ever seen. This means that your brother and his dog have made it back home to Heaven," said *Emilio*. "That's why God, in His infinite wisdom, has smiled to us, making this Star even brighter."

Hearing this, I was sure glad that Shep had spent so much time with me, teaching me how to hunt and see life. I named that star Dog Star. But, when I tried to tell my father and mother about what *Rosa* and *Emilio* had told me about Shep running around the house and then intercepting Joseph's soul to guide him back to Heaven so that my parents could feel better, my dad and mother got angry. I couldn't figure out what was going on. I was saying something to help them.

"Your mother and I got the finest doctors that money could buy for your brother!" yelled my father at me. "And no stupid old Indian beliefs are going to CREATE DOUBT IN OUR WORD!"

"But *papa,*" I said, "I'm not trying to create doubt in your word. I'm just trying to tell you that Joseph and Shep have made it back to Heaven.

"STOP IT!" he yelled.

"But *papa,*" I said, continuing to speak, "please just listen to me, even *mama,* she has told me that her mother always said that we're walking stars coming to Earth to do our work, then we return to Heaven to help *Papito* keep the stars bright. So don't you see, for Shep to guide *Chavaboy's* soul back up to Heaven is only natural," I added.

But you would've thought I'd said something terrible, because my mother ran out of the room, yelling that she didn't want to hear anymore!

"See what you've done," said my father. "Open your eyes. Eh, can't you see, *mijito,* your brother is dead! HE'S GONE! THOSE ARE JUST BACKWARD INDIAN SUPERSTITIONS to give people FALSE HOPES! You're mother is right! We live in this country now! We can't have all those old ways from Mexico anymore!"

"But . . . but I don't understand? Why not, *papa?* Even Joseph told me that what is really needed in the world isn't control or money or new inventions, that what's needed is—"

"BECAUSE, DAMNIT!" shouted my father with such power that the cords of his bull neck now came up as big around as ropes, "WE GOT TO KNOW OUR PLACE!"

Hearing this, my heart leaped! Now I, too, was scared. My very own *papa* was now telling me the same thing that the kids and principal had told me at school about me needing to know "my place."

"So, then, what is our place, *papa?*" I asked, my heart beat, beat, beating with terror!

"DAMNIT!" he screamed again, his face twisting with confusion. "I DON'T KNOW!"

And here, my dad, my hero, the strongest, toughest man in all the world, began crying, too, weeping like a child. Tears came to my eyes. And a cloud came overhead, taking away the sunlight. The day turned cold. And I could see that my brother had been very smart to leave. I wished that I'd gone with him.

That same day, my mother and father decided to fire *Rosa* and *Emilio,* because they said they couldn't have people on the ranch telling my little sister Linda and me about the old ways of Mexico and undermining their authority.

Linda and I started crying. We didn't want *Rosa* and *Emilio* to leave. We loved them. They'd become like our second parents. Besides, firing them wouldn't change anything for me. Because my sister and I had both seen what we'd seen, our brother's dog Shep had really, really gone crazy-*loco* with his *amor* for our brother and then disappeared the morning Joseph died, never to be seen again.

Our *tía Tota* could talk all she wanted about her silverware and our parents could talk all they wanted about the best doctors money could buy, but I still knew what I knew: animals had Sacred Souls and we, humans, had to pay close attention to them, so we could then get back into Heaven, with them guiding us.

BOOK three

I was CRUSHED! BEAT! Nothing worse could ever happen to me! But then, it was only a few days after my brother Joseph's death, and I was told by our teacher in front of everyone in my class that I wouldn't be passing to the fourth grade once again.

I began to cry. I just couldn't help it. I was tired, and so beaten down that I now even had trouble breathing. Seeing me cry, our teacher told me not to worry, that this time I wasn't alone, that Jeffrey, another student, was also not going to pass. I looked at her. What was she, stupid or constipated, as my dad always said of people who didn't think straight? Did she really think that it was supposed to make me feel better that someone else was flunking along with me?

I turned around in my seat to look at Jeffrey and saw how he was taking this announcement. It was strange, I'd never noticed this kid before, but now that we'd both been told that we were going to flunk, I could remember some very interesting things about Jeffrey.

One day, about a month back, when our teacher had been reading to us about the local Indians, some of the kids had said that they had been so backward that they hadn't even had the brains to put any windows in their huts so that they could look out. Jeffrey, who was always real quiet, spoke up.

"They were probably more interested in keeping warm," he said, "instead of looking out. And besides, because of the local weather, they probably only went inside for sleeping and did all their cooking and eating and living outside."

Our teacher, I'll never forget, looked at Jeffrey as if she'd never seen him before. Then she'd said, "Isn't it interesting that our second slowest learner in the class is the one to think of this. I guess that maybe there's a place for all of us."

Boy, was I ever shocked that our teacher had called Jeffrey "the second slowest learner" in front of everyone. Hell, I'd never even realized that he was slow. I'd automatically turned to look around, wondering who was the slowest learner of all, and then I noticed that a lot of the kids were looking at me.

Suddenly, it entered my brain that I was the slowest learner in the class. I'd felt like sliding down in my seat and disappearing. I felt like dying—I felt so ashamed. But then, I heard that little voice deep in the back of my head tell me to remember that the teacher also said that there was maybe a place for all of us. I felt a little better, because maybe this meant that there was a "place" even for me, the slowest learner, and maybe even a place for my dad and mother, too, if we could just find out where this place for Mexicans was, then everything would be okay.

At recess, after the announcement of Jeffrey and me not passing, I went over to try and befriend him so maybe I could make him feel better, because, after all, I'd already flunked once, so I had a lot of flunking experience.

But when I approached Jeffrey, I saw that he didn't want to be seen with me. In fact, he quickly glanced around and moved away from me. I didn't trail after him. I knew what was going on. He was still trying to fit in with the other kids like I'd tried to do my first year in the third grade, plus he was also White, so he didn't want to be seen with another flunker, who was a Mexican, on top of that.

My heart broke, I felt so rejected. Quickly, I walked back across the asphalt by myself. I decided to sit as far away from everyone as possible. This way I figured no one else could reject me. Sitting alone, I watched the other kids, and I could see that it had worked for Jeffrey. Now that the other kids had seen how he'd shunned me, they were accepting him. My teacher came over and asked me why I was alone.

"Why don't you play with the other kids?" she asked.

I felt like spitting on her. Why had she branded me a flunker and the slowest learner in front of everyone if she really wanted me to be accepted by the other kids? But I didn't spit on her as I should've. No, I just shrugged my shoulders and acted like I was too stupid to know what was going on. Acting stupid, I was beginning to find, was a good way of avoiding further embarrassment.

But things really weren't so bad. There were still two or three kids who'd speak to me. One was named Phil and he lived by the lagoon south of school overlooking the water. And the other was Billy, who lived up the hill from our place where California Street dead-ended. Both Billy and Phil had older parents, and they came to school clean and well dressed.

When I brought the note home, explaining to my parents that I'd flunked the third grade again, my mother started crying, not knowing what else to do. But my dad knew what to do. He took me outside and over across the grass to the large old pepper tree.

"How old are you?" he asked.

"Nine," I said. "I just turned nine a couple of weeks before . . . before Joseph—" I stopped my words. I didn't want to upset my father.

"Died," said my dad. "You turned nine a few weeks before your brother Joseph died. Right?"

"Yes."

"Then go ahead and say the word. We can't go on with our life if we're afraid to say that word."

"Died," I said, tears coming to my eyes. Suddenly a breeze came up and the limbs of the big pepper tree began to sing.

"Good," he said. "You know, I guess that your mother and I haven't been paying too much attention to you and your sisters with your brother's sickness and death, have we?"

I shrugged. I didn't know what to say.

"I'm sorry," he said. "Damnit, my father did the same thing to us, and here I end up doing the same thing with you kids." He took a deep breath. "My dad, he didn't even have the eyes to see me after all of his tall, blue-eyed sons were dead. He started bellowing from the hilltop to hilltop like a madman, saying that God had forsaken him, and he had nothing more to live for, because all his sons were gone!" Tears streamed down my dad's face. "I was about nine or ten, the same age as you, but he couldn't see me, because I was dark and Indian-looking and didn't have blue eyes like him. Your mother and me, we are not going to forget you and your sisters like my father forgot me. I swear it!"

And saying this, he took me in his arms and held me close, kissing me and hugging me over and over again, until he finally stopped crying. Drying his eyes, my dad pushed me away, held me at arm's length, and looked at me eye to eye.

"Do you remember me telling you what my mother did once she saw our father reject us kids who were still alive, and he start drinking himself to death, grieving over all his dead blue-eyed sons? My mother, nothing but a little bag of Indian bones, saw this same reality, that they'd lost eleven more children, but she didn't fall apart and say that God had forsaken her. No, she said, 'I got three children left to live for,' then she added the most powerful Indian saying that we got in all Mexico, '*mañana es otro milagro de Dios*', tomorrow is another miracle of God's, and with these words, with this conviction of heart and soul, I saw her rise up with the power of a mighty star and take us off the highlands of *Jalisco* and down through the valley of *Guanajuato* where we saw whole towns with dead bodies piled up high. And that little old woman never gave up!

"We are not going to panic and lose faith," he added, holding me by my shoulders, "because of your brother Joseph's death, and stop living. Do you understand? We are *una familia* and we are going to keep on living with *amor* in our *corazones*, because truly, *mañana es otro milagro de Dios*, and who knows, maybe it was best for your brother to pass over like this at an early age. He was so goodhearted and thoughtful, that maybe the world would've . . . well, broken him. He wasn't a real *cabrón* like you and me. Do you understand?"

I shrugged, but then I nodded, because maybe my dad was right and I was a real *cabrón*. After all, I blackmailed people, and I'd paid a girl five cents to see her underwear. And I'd also bit my teacher. "Maybe," I finally said. "I think I do."

My father was looking at me and had the biggest grin I'd ever seen. "Do you know what you just did?" I shook my head. "You just closed your eyes and wrinkled up your face so you could think, just like my mother used to do."

"I did?"

"Yes, like I always say, blood knows blood, and your brother was a lot like my older brother José, and you're a lot like my mother."

"You mean that blood can also cross and go from being male one time, to then being female the next?"

He nodded. "Sure. Why not? Isn't this what God, Himself, does, going back and forth between day and night, male and female."

I shrugged. I didn't know.

"Hell, I'm glad that you flunked again," said my dad.

I was shocked. "You are?" I said.

"Hell, yes! My mother always used to say that when the going got tough, this was when we had the real opportunity to do God's work. 'Come on, God,' she'd say in the middle of disaster, 'give it to me some more! Because I know that together, You and me, God, We can move mountains!' And why did my mother—bless her soul—talk like this, because, *mijito,* she didn't believe in God. No, she LIVED with GOD! Get it? She saw herself as Living with *Papito Dios,* and He needing her just as much as she needed Him, so that His Will will be done!

"Shit, if you had done too good in school, you might have ended up wanting to become a teacher. And you look at these teachers, most of the ones that I've seen look so scared, that they'd never stand up for themselves. Your principal, could he ever survive in prison? Would he know what to do in Las Vegas with a thousand dollars on the table? Could he land in Mexico with no money, speak no Spanish, and build a life from nothing? He wouldn't even have the guts, I tell you, to step up to the high rollers' table, or go to Mexico broke."

"*Papa,* will I have to go to prison?" I asked.

"What? Why do you ask this?"

My heart was beat, beat, beating! "Because of what you just said about the principal. You keep saying that, well, to be a man we must know how to survive in prison."

When he saw my fear, his eyes softened. "No, *mijito,*" he said, "you might not end up going to prison. But you are *un Mexicano,* and so if you don't kiss the ass of authority, or learn to be real cunning, then you might end up needing to do some time in prison."

My heart was going crazy-*loco!* "Would have *Chavaboy* gone to prison if he had lived?" I asked, almost pissing in my pants, I was feeling so scared. I didn't want to have to go to prison. I was still too little. I wanted to stay with my *mama* and *papa.*

"No," said my father, "I don't think he would've. *Chavaboy* was different than you and me. He was more like your mother. You're more me, a real *cabrón.* Remember how you pissed on me when you were still in diapers," he said, laughing. "And now you blackmail people."

I froze. I didn't know that my dad knew about my blackmailing. "How do you know about that?"

"What? About your blackmailing?" he asked.

I nodded. I was sure that I was now in deep trouble. But my dad only laughed.

"Don't be frightened," he said. "Remember, fear is only good when the other guy's got it and you don't."

"Then you're not mad at me for blackmailing people?"

"Hell, no! I love it! But also, you're now too big to keep doing that, because . . . blackmail is a serious business that can get very dangerous. I've seen grown men, tough men, get killed because they didn't know when to quit. A man needs to know when to quit, *mijo.* This is the secret—not just to gambling and blackmail—but to all of life."

"Quitting? But I thought you told me that a good man never quits. That he's a *burro macho,* and just keeps going and going, no matter what."

My dad smiled. "Good. Wonderful. You've just come to the crucifixion of life, *la vida.* You see, in everything we say that's true, the opposite is also true. This is how we catch our balance and have understanding, instead of opinions. Opinions, you see, they're like assholes, everyone's got one, and they all stink. But understanding, she smells sweet, and she's the start of wisdom. *Capiche?*"

"I don't know," I said, shrugging. "Maybe I do, a little bit."

"Good. The beginning of all wisdom is to understand that you don't know. To know is the enemy of all learning. To be sure is the enemy of all wisdom. Look, yes, a good man never, never quits when the going gets tough, or great tragedies hit your *familia*. But also, a good man *a las todas* has got to not be stubborn and know when to say no to that next drink, or to that next turn of the cards, and walk away from the game. Your mother, she's the one who, with all her power, finally got me to quit my wild gambling and bootlegging, and go legal.

"No, we good men never quit. Never! And yet we got to know when to stop and say '*mañana es otro milagro de Dios*' and get a good night's rest so we can see things a little more clear in the morning."

I nodded. This made a lot of damn—I mean, blessed—good horse sense to me. Then it hit me like a lightning bolt; horses did think, and a lot of people knew this, or then this expression of "good horse sense" wouldn't even exist! I suddenly felt very good. This was when the humming began again behind my left ear. I smiled. *Papito Dios* was massaging my head. He was telling me that He agreed with what I'd just figured out.

"Are you feeling better?" asked my father.

"Yes," I said, "I'm feeling a lot better, *papa.*"

"Good," and saying this, he gently took me by the shoulders and turned around, and we headed back home. The branches of the whole huge old pepper tree began to dance in the breeze. I just knew that my dad and I were not alone.

The next morning my dad surprised me again by telling me that I didn't need to go to school if I didn't want to. "Over is over," he said, "so why the hell go to school, eh? What are they going to do, flunk you twice?" he said, grinning. "I say we all eat breakfast together this morning and then we all go horseback riding like *una familia de LOS ALTOS DE JALISCO!*

I'll never forget that morning, I felt so good to eat *huevos rancheros* with my *familia,* instead of just eating alone with my father so I'd have time to do my chores before I went to school. And at the corrals, after we'd saddled up, my father reached in his pocket and gave me five single dollar bills. "From now on, no more blackmail, but I do want you to start carrying money in your pocket at all times, *mijo,* because money is good, and not just to spend on candy and *pendejadas* like that. But also to give to our people who come up through our valley, following the railroad tracks below our home. Some of these poor *gente* come all the way from deep in *Mexico* by foot, like my mother and I did.

"They're tired, hungry, and looking for work. They're good people, *mijo,* the best! Not wanting nothing for nothing. You give them money if they're broke so they can go to the store to buy bread and milk and meat. This is the great miracle of money, it can save a hungry person's life. I know, we were starving to death when we crossed the border. Hunger is what I've feared all my life; not death. Dying is easy compared to being *un pobre mendigo* that can't speak the language, not even to ask for water.

"And when you use this money up, come ask me for more, or just go to our room and get some more out of my pants pocket, then tell me how much you took. You're nine years old, a man of honor, and your honor has no price tag that *dinero* can ever buy. Do you understand, you got money, you know how to handle a gun, and no man or woman, no matter how big or important they think they are, can ever walk on your shadow again without your permission. Got it? You are now *UN HOMBRE,* and to be respected!"

I nodded. I got it. My heart was beat, beat, BEATING with *AMOR!* I took the five dollars and put it in my pocket, feeling like I'd grown up ten years since I'd been told that I'd flunked again.

That morning my mother and little sister—Linda hadn't started school yet—went horseback riding with my dad and me. We went out the front gates, down Stewart Street towards Cassidy, turned left, went up the hill, then cut down the slope by Dr. Hoskins's place towards the duck lagoon between Oceanside and Carlsbad.

"I hope that man burns in hell," said my mother as we rode by the doctor's place.

"Lupe," said our father, "*cálmate*. We're out riding and having a good time."

She made a face at our dad. "You're a man, Salvador!" said our mother. "Not a woman! You'll never understand what a mother feels down here in her heart when she loses a child so young that he never had the chance to live!"

She burst into tears. Our father took a big breath, but he said nothing, and we rode across the hill towards the east, and it was beautiful. We were now riding on the big ranch east of *El Camino* that had just been bought by a world-famous Olympic champion ice-skater. From way up here we could see down the whole lagoon all the way to the glistening Pacific Ocean in the distance.

We got off our horses, loosened their cinches, and let them graze for a while. My father brought out his pint bottle of whiskey, took a swig, then offered it to our mother. I'd never seen our mother drink like this before, but she did. Still, when my dad went to kiss her, she turned her face away. Seeing this, I once more flashed on what our *tía Tota* had said about not wanting kids, and I wondered if our mother was now, also, maybe sorry that she'd ever brought any of us into the world.

My father said nothing to our mother about her turning her face away from him, and we cinched up our saddles again. Then my dad helped my sister and me get back on our horses, then he helped our mother, too, stroking her leg. After that he leaped, without putting his foot in the stirrup, right up on top of Lady, his big Morgan mare.

Getting home, we unsaddled, ate lunch with plenty of big, juicy avocados, freshly made cheese, and homemade *tortillas* and *salsa*. That afternoon, I went with my parents to take a lug of avocados to my teacher. And strangely enough, my teacher didn't look so big and strong to me anymore. No, she looked kind of old and weak and nervous to see me come in with my parents. My dad and mom gave her the box of avocados. The lug had an envelope of money in it, along with the avocados. Seeing the envelope, my teacher didn't want to accept the box.

"Oh, yes, you do," said my father in a firm but calm tone of voice. "You see, we got a little plan. Lupe called our priest and spoke to him this morning, see."

My mother then told my teacher their plan. My teacher was to pass me, and then I wouldn't be coming back to the public schools anymore. Instead I'd start attending Catholic school, which my dad had already explained to me was a much better and more dependable organization where money didn't just talk, but could actually scream, and get you a guaranteed passage—not just of grades—but into Heaven, itself, with a fat-enough envelope.

My teacher saw the light, passed me, put it in writing, as my mother suggested, accepted the avocados and the envelope, and we went home whistling a happy tune. Boy, I loved this! I'd never realized that life could be this easy! Blackmail and bribery were really the way to go. But also, I realized that I'd promised I wouldn't do blackmail anymore. Nothing had been said about bribery.

I was sound asleep when my mother came rushing into my room, grabbed me, started shaking me, telling me to wake up!

"HURRY! Get dressed!" shouted my mother. "Your father has gotten his horse drunk in a bar again!"

I only had a couple of days of school left. My father had just purchased a new horse named Cherokee. He was a white palomino that my dad had gotten in a trade when he and Jimmy Williams sold Blue Eyes to Harold Figstad, who'd opened up the new Mobil station on Hill Street and Cassidy.

"Which horse?" I asked my mother. I hoped it was Lady, his old Morgan mare, because Lady knew how to hold her liquor and behave herself. When my dad got to drinking, he sometimes also liked to give his horse a few beers. Luckily, my dad never gave his horses hard liquor. Only beer. But Cherokee, he was a young gelding who'd just been castrated, and so he probably had no idea how to handle a few beers.

"I don't know!" said my mother. "I didn't ask. I guess it's his new horse Cherokee. All I know is that the horse has gotten wild and broken furniture and nobody knows how to get him out."

Boy, was I ever sorry to hear this. This could, then, really be bad, because a horse that didn't know how to hold its drinks, was a very dangerous animal.

"Hurry! Get dressed," said my mother. "I'll get your father home, and you ride the horse."

I was ready to pee in my pajamas, I was so scared. I hoped the bar was the Pepper Tree that had a dirt road behind it, because I didn't want to have to ride a drunken horse on concrete or slippery asphalt.

I quickly pulled off my pajamas, got on my Levi's, my socks and boots, too, then put on a long-sleeve shirt, washed my face, and grabbed my spurs and cowboy hat. I didn't know if I should wear spurs or not. It all depended on the horse being willing to go or not.

I ran down the stairs, out the front door, past the patio, and got in the car where my mother already had the motor going. We quickly drove down the long driveway, out the big white gates, turned right on California Street, went past old man Hightower's new market, and turned left of Hill Street. About a block and a half down the street, my mother pulled over and parked across the street from Figstad's new Mobile station. The great big red Mobil horse with flying wings was blinking off and on. We got out of the car and went into a place I'd never noticed before. It had a huge double-door entrance, but there was a solid wall right in front of the doorway, so that you immediately had to make a sharp turn to the right or to the left in order to get inside. I could hear my father's big booming voice before we'd even gotten through the entrance. I could hear tables and chairs being knocked around, too.

Following my mother in through the right side, we rushed into a large room and there was my father by the bar with two women at his side, and his golden-white horse was slipping and sliding on the dance floor, trembling with fear and knocking furniture all over the place. My father had his Western hat pushed back on his head and he was singing in Spanish like *Jorge Negrete,* with a bottle of beer in his hand. He didn't seem very concerned about Cherokee tearing the place up. Seeing my mother and me, his whole face lit up *con gusto.*

"QUERIDA!" he shouted as the two women quickly moved away from him. *"ALMA DE MI CORAZÓN!"* he bellowed. "You've come to join me, at long last!"

"No, Salvador! I've come to take you home!"

"Home? But, why home? The party is just beginning!"

I could see that all the other people in the bar were staying way back, too. Obviously, they were all afraid of the horse and all his kicking and slipping and sliding. There were more than a couple dozen people in the place, about half men and half women. The bartender—I thought I recognized him, but I wasn't sure. He, too, looked very nervous. I guessed that he was the one who'd called my mother, instead of calling the cops. Which was good, because most of the cops wouldn't have known what to do with Cherokee. Last year, two young cops had tried herding some of our cattle home that had gotten loose and ended up on the beach, and one cop had gotten his motorcycle too close to the cattle and the headlight of his motorcycle had gotten kicked out.

I quickly went over to Cherokee and picked up his reins, which he was stepping on, and pulled him up close, so he'd realized I had him.

"Easy, big boy," I said to him. "Easy does it." But I could see that he was so drunk and scared that he didn't know what was what.

"LUPE, *MI CORAZÓN! Mi novia! MI ESPOSA!"* my father kept shouting. "LET'S DANCE, YOU AND ME, *MI AMOR!"*

"Salvador, you're drunk!"

"So what! I still got feeling for you as big and deep and strong as the first day I lay eyes on you! You are MY ANGEL!" he bellowed, reaching for her and trying to kiss her.

"No, Salvador, please! You stink!"

"Yes! I stink LIKE A BULL who still breathes and adores every step you take, shifting weight from one beautiful hip to another beautiful hip with a poetry of movement that puts my HEART ON FIRE!

"He died! He's gone! And I, too, loved our son with all my heart, but what the hell are we to do? We're still here and breathing! Come, let's make the kiss, *querida.*"

"No, Salvador!"

"Yes, Lupe!"

"NO!"

"YES!"

I was trying to lead Cherokee out the crooked entrance, around the wall that was in front, but he panicked and reared up—almost hitting his head on the ceiling—then slipped on the floor and fell back on his ass. He looked like a huge puppy dog sitting on the dance floor.

I laughed, but people screamed, scrambling away, not thinking it was funny in the least. I held on to the reins as best I could, as Cherokee came leaping, slipping up on his four hooves. My father was instantly at my side.

"Hold him, *mijo!* You're a *charro de Jalisco, a lo CHINGÓN!* Ah, life is so full of wild, twisted adventures, eh?" he said, with a drunken grin, licking his lips with his huge tongue. "You got troubles with a horse, I got troubles with THE WOMAN I LOVE!

"So the most important advice I can give you about life, *mijo,* is this . . . you lift up the tail, stick in your nose, take a big sniff, and you'll always know who is a female! Do you hear me! LIFT UP THE TAIL, stick in you nose, JUST LIKE A STUD! And take a big sniff, THE BIGGEST! And you'll ALWAYS KNOW WHO'S A FEMALE!" he repeated, licking his lips with his long, thick tongue.

"Look at your mother over there! She's so BEAUTIFUL! She just comes into the room, and I start smelling! And what is it that I smell, it is the SMELL OF LIFE, *LA VIDA,* WITH ALL HER LUST AND BEAUTY—so RICH A SMELL that it brings tears to my eyes, blood to my *corazón,* and strength to my *tanates!*"

"SALVADOR! STOP IT!" yelled my mother, coming to us. "You're an embarrassment to everyone!"

"Well, then, TO HELL WITH EVERYONE! Because, how can we stop it? Eh, you tell me, Lupe. Wars, starvation, death, the whole damn catastrophe of life happens to us, generation after generation, and yet nothing can stop this smell *de la vida* that WE SMELL!

"Hell, a bull smells a cow in heat five miles away, and that bull tears down fences to get to the cow, BELLOWING THE WHOLE WAY! And women, thank God, once they've opened their hearts, they're in heat every day of their lives, forcing us men to want to live! TO LIVE SO WE CAN LIFT THE TAIL, STICK IN OUR NOSES, and take a big—"

My mother slapped my father before he could finish his words.

Everyone was staring at my parents, especially all of the women. They had a liquored-up, wild, hungry look.

"SLAP ME AGAIN!" he shouted. "At least this is better than not touching me since our boy died! You tell me that a man can't feel what a women feels! You tell me that a father can never understand what a mother feels when they lose a son! Well, I say BULLSHIT! AND COW SHIT, TOO!

"My mother, God bless her soul, lost child after child in the Revolution, but it was not she who broke! It was my father who fell apart, saying that God had forsaken them, and it was THE END OF THE WORLD!

"But *mi mama,*" said my father, staggering towards my mother with tears streaming down his face, "she never said that! No, she said '*mañana es otro milagro de Dios,*' and with that conviction, she took us off the mountains of *Jalisco* and we migrated to Texas border. And that old Indian woman was a MOTHER! And she never, never, never once gave up on life!

"Come to me, Lupe! Come to these waiting arms! I love you *con TODO MI CORAZÓN!* The touch of your skin, the smell of your—"

"SHUT UP! SALVADOR! For the sake of God!"

"Okay, I'll shut up! BUT I NEED YOU! I NEED TO HOLD YOU CLOSE! For how can we keep faith in life, if we don't have the smell and warmth of each other to keep us going. I LOVE YOU!"

"Come, Salvador," said my mother, glancing around at everyone. "We need to go home . . . *Mundo* will ride your horse."

"Will you hold me, skin to skin?" he said, rocking back and forth on his feet. "Will you love me like I love you? Will you let GOD DO GOD'S WORK, and . . . and . . . Lupe! LUPE! LUPE!" he bellowed, dropping to his knees with open arms. "*TU ERES MI ESPOSA!* The handcuffs of my heart and soul!"

She didn't want to, she really didn't, but she then went to him with open arms, too, and they drew each other close, held, and then they began to kiss. He was kissing her, and she was kissing him. Not one single sound could be heard through out the whole bar. Everyone stared at them, mouth open. Even the horse seemed to have calmed down.

My dad stood up and they continued kissing, making that kiss that my father had been asking to make for weeks. Then finally, my mother and father turned and went out the front doors, and I was left with the drunken horse. But every time I tried to lead Cherokee to the front door, he'd rear back in panic. Finally, it dawned on me that something very bad must've happened to him at the entrance.

"Hey!" I yelled at everyone. "Did something happen to this horse at the entrance?"

"Yes," said the bartender, "he hit his head on the doorway and got knocked down to his knees when your dad rode him in."

"Oh, then that's why he's scared of the entrance. Is there any other way to get him out of here?"

"I don't know anything about horses," said the bartender. "But I'd guess that the back door isn't big enough for him to fit through."

"Can somebody hold the horse, while I go take a look?"

But no one would come forward to help me, until one of the women who'd been near my dad when we'd first come in, now walked around from the far side of the bar. She had red hair and was older, but very youngish-looking and beautiful.

"I'll hold him," she said in a deep, throaty voice.

"Do I know you?" I asked.

"Yes," she said, acting a little nervous. "I've come over riding to your place a few times."

"Oh, I see," I said.

She took the reins and spoke gently to the horse. I could see that she knew what she was doing. The bartender took me to the back. We walked past cases of beer and wine and whiskey. Getting to the back door, I could see that he was right—the door was way too narrow. The front entrance was still the best way to go. We went back to the front of the bar. I hadn't noticed all the cigarette smoke before. The place was a sea of gray-white smoke and the horse was now heaving like he was going to get sick. The youngish older woman could barely hold him, he was slipping and sliding so much.

"How much did he have to drink?" I asked.

The bartender looked embarrassed. "About six beers and a couple of *tequilas.*"

"Oh, my God!" I said. "One shot of *tequila* alone will kill a five-hundred-pound pig."

"Well, what could I do? Nobody can tell your father no. Hell, he said he'd buy the place just to fire me, if I didn't serve his horse."

I believed him, and decided to put on my spurs. This way I'd just lunge the horse through the entrance, and hope to God that I didn't get knocked off the saddle by the top of the doorway. It looked to me like there were only a few inches between the saddle horn and the top of the doorway. How my dad had ridden the horse in, I couldn't figure out.

Getting my spurs on, I told everyone to get back, which they all quickly did, except for the red-haired woman.

"I'll hold him steady for you while you mount," she said, which made a lot of sense, the way the horse kept slipping all over the place.

The whole bar was in shambles with all the havoc that Cherokee had caused.

"Get him near that table," I told her. "I'm still too little to get on a horse without a fence."

"How about a bar stool?" asked the barkeep.

"That would be real good," I said.

He brought a bar stool over to me, but then backed away as fast as he could. I guess that they'd all seen Cherokee break things up, and were very wary.

I got the stool and took it over to the nervous young gelding, talking gently to him the whole time. Then I checked the cinch, which was a good thing, because it was loose. I cinched him up good, tried to shorten the stirrups, but saw that they'd never go up high enough for my short little legs.

"To hell with it," I said, and climbed the bar stool. I was just about ready to put my foot in the stirrup and scramble up into the saddle when I felt two strong hands grip me, pick me up like a feather, and put me on the horse. I turned around to say "thank you" but I saw no one was behind me. Then I smelled my brother. I could smell that maroon woolen robe that he'd been using the last year of his life.

A great calm washed over me, realizing that my brother Joseph was here with me. And it was a damn good—I mean, blessed good thing, too, because the next thing I knew, Cherokee reared up, wanting to go!

I looked at the big plate-glass window that was right alongside the front entrance and wondered if I wasn't better off just leaping Cherokee through it. But then, I remembered that Bert Lawrence, our horse-shoer, had done that last year at the Mira Mar restaurant at the north end of Oceanside after the Fourth of July fireworks, and broke his horse's leg. I decided not to try the window, which looked pretty tempting, and I pointed Cherokee towards the front entrance, and when he started to panic, I suddenly gave him the spurs with everything I had.

The horse bolted like a bat out of hell, slipping and sliding, and we shuffled around that wall and out the front door, with me bent over in the saddle real low so I wouldn't get ripped off the horse by the roof over the entrance.

Cherokee fell to his knees as we went through the doorway, skidded, scrambled, got back on his feet, and leaped forward.

A cold wall of fresh air belted us as we hit the street, and there were speeding trucks and cars with their bright lights right in front of us. But this didn't seem to frighten Cherokee. No, what he did next showed me that he was as quick and agile as any horse I'd ever ridden, and brave, too. Because the son of a bitch didn't hesitate a pregnant-frog's ass to leap over the first car, then dodged a truck, and there was the red Mobil horse blinking off and on right in front of us, and shit, WE WERE NOW FLYING, TOO!

Hooves echoing at a gallop, we went right down the middle of Hill Street with cars honking and braking, and at the end of the first block, we turned right, hit dirt, and I now gave the horse the spurs again. I was beginning to understand that the faster we went, the better Cherokee could keep his drunken balance.

We were racing up the street, flying past houses and neighborhood dogs. The Oxely girls, older than me, were coming in from a date, I guess, and I could see that they couldn't believe what they were seeing—a drunken horse racing by them in a wild, wobbly run.

By the time I got to the gates of our *rancho grande,* I realized that a car had been following me the whole way. It was that same red-headed woman who'd helped me in the bar, along with another woman, too, and once they saw me go in through the big white gates of our *rancho grande,* they turned around.

Cherokee was pouring with sweat and beginning to sober up by the time we got to the corrals. It was a good thing my feet hadn't been able to reach the stirrups, because the moment we quit running, he fell out from under me, rolling over on his side. I scrambled away as fast as I could so I wouldn't get caught under his eleven hundred pound body.

He was sick. He was gagging. I guess it must've been his first drunk, because he really didn't know how to handle his liquor like Lady. Hell, the old Morgan mare would've just let me lead her out the entrance of the bar and walk her home as easy as eating apple pie.

The next morning, my parents didn't get up for breakfast and they barely made it to lunch. When my mother did get up, she fed her birds, and was whistling like one of her canaries. I guess that my parents were in love once again.

CHAPTER **eighteen**

Waking up the next day, I just knew that something kind of extraordinary had happened to me last night when I'd ridden that drunken horse home. It was like I now just knew deep inside of myself that I wasn't going to die like my brother had died. I could now see so clearly that my brother Joseph had been absolutely right when he'd said that "his" was very different from mine, that "his" had been going on for a long time and so he'd already used up all of his nine lives, but I, on the other hand, had probably only used two or three of mine.

Racing through the streets on Cherokee, I'd also come to know deep inside of me that I wasn't alone, that none of us were ever alone, especially when we had so much *familia* up in Heaven pulling for us. I had my brother Joseph in Heaven. I had my two grandmothers. I had my father's older brother, *José*, the Great, and I had my great-grandfather, *Don Pio*, the greatest man who'd lived next to *Benito Juárez*, the Abraham Lincoln of Mexico! Heaven was full of my *familia* and our *amigos*, too, and so this was why Cherokee hadn't tripped and fallen face first in the street into oncoming traffic. Strings were being pulled for me up in Heaven, lucky strings!

Going to school that day, I don't know how to explain it. I now felt so much older than the other kids in my grade. And I was older by a year, because I'd flunked the year before, but there was something else. No, I mean, I was way older somehow. In fact, the teacher, herself, and even the principal didn't impress me or scare me anymore. I was now *un hombre,* and no one was ever going to walk on my shadow again without my permission, as *mi papa* had so well explained to me.

People now had to speak to me with respect and they had to make good horse sense, too. They could never again bully me just because they were older, bigger, were a priest or a teacher. And man, did this feel good deep inside of me, especially after we, the Mexican kids, had gotten beaten down so much ever since kindergarten. And that playground teacher, the *gallo-gallina,* as we'd nicknamed her, I sure wished that I knew where she lived. She'd tortured us worse than anyone and she'd enjoyed it. I'd like to set a dynamite charge to her house and blow her ass to smithereens.

God must've heard my request, because the very next day, I heard my mother call information and get someone's phone number and their address. This meant that all I had to do was pick up the phone and call information to get that damn *gallo-gallina's* phone number and home address, so I could have my revenge if I really wanted to. I loved it!

Then school was out! I was free for the summer. And my parents were still behaving like they were in love. They decided to take a trip to Mexico for a month with our older sister Tencha. My sister Linda and I were not going to go. We were to stay with Hans and Helen on their chicken ranch in Bonsall. I hated to do this, because I wasn't going to be allowed to take the .22 Winchester that my dad had just bought me, so I could continue to practice my shooting every afternoon. And I was getting really good with my pump-action .22 Winchester.

Our parents left with Tencha in our new Cadillac, and my little sister and I went out to Bonsall. My sister and I learned German and how to make our beds with square corners and how to clean the eggs and feed the chickens and mark the chart of each chicken, so that we'd know who laid eggs and who hadn't. After two weeks of not laying regularly, these nonproductive chickens were beheaded and cooked for dog food. Uncooked chickens were never fed to the ranch dogs, Hans explained to my sister and me, because the taste of fresh blood could turn even a good dog into a chicken killer over night.

A few days later, I said that I wanted to have guitar lessons. That song of my brother's with the cowboy chasing the Devil's red-eyed herd just kept singing in my head all the time. Hans started driving me to

Oceanside twice a week so I could get guitar lessons. I told my guitar teacher that I wanted to learn the words to my brother's song. He got me the sheet music, and my sister and I began to recite the words to the music every afternoon. I don't know how, but Linda could already read better than me and she hadn't even started school. Singing "Ghost Riders in the Sky" under the big pine tree felt so good and full of magic, and yet it also felt real sad.

When our parents returned, I didn't recognize them at first, especially with the funny, colorful, old-world Mexican dresses that my sister and mother were wearing. That evening, my mother and Tencha showed us all the pictures that they'd taken in Mexico, especially in Acapulco.

The summer ended and the days began to shorten, and it was almost time for me to start school again. This year I'd be going to the Catholic school at the San Luis Rey Mission. But I didn't want to go to school. What was the point, to just find out that I was stupider than the year before? I could very well see that every year that I went to school, the reading was getting harder and harder, and I was just getting left behind further and further.

I took my sister Linda aside—because this year she was starting kindergarten—and explained to her it was not a "kinder garden" that she was going to go to, as our parents thought. That teachers could be mean, especially since we were Mexican. She laughed and said that she'd just tell them that she was Chinese, if they asked her, because she loved Chinese food, like our father. I told her that I didn't know if that would work. But she seemed so sure, that I just didn't say anything more.

Late one afternoon, just a couple of days before my sister Linda and I were to start school, I suddenly felt that little quiet purring behind my left ear, and I instantly knew that I was supposed to saddle up Midnight Duke and go riding down to the sea.

Then I remembered that this was exactly what my brother Joseph had wanted to do before he died. I hurried over to the horse stables, caught Duke, haltered him, took him to the fence by the ramp where we loaded the cattle into our cattle truck, brushed him out real good, then I hauled my blanket and my little saddle up on the ramp and got myself up high enough—because I was still too short—so I could saddle up. Then I climbed up on the fence, took off the halter, and bridled Duke, got off the fence, turned him about, cinched him up good one last time, then climbed back up on the fence so I could mount him.

It was really late now. This was truly not a good time for me to go riding down through the marshlands to the sea. I'd seen how much trouble my brother had gotten into for taking just one wrong turn. But for some reason, I wasn't scared. No, I was all excited inside. It was like this purring, this humming, was guiding me just as I'd been guided home off that tall *mesa* out by the cemetery where Shep had leaped into the sky to intercept my brother's soul, and I'd also been guided the night that I'd had to ride Cherokee home from the bar.

I mounted and quickly took off, riding fast so I could get across the marshes while there was still plenty of light. How I'd get back in the dark, I had no idea, but I just knew that everything was going to be okay. After all, *Chavaboy* was up in Heaven, so if I needed any help, he could get hold of our two grandmothers and Sam and Shep and even Jesus and Mary, if need be. I mean, the Heavens were now full of human and animal souls just ready to help me out.

Quickly, Duke and I cut through the flat marshes as fast as we could, running when we could on the solid ground, then nibbling our way real slow and easy when we had to. Then we were at Hill Street, the coastal highway at the west end of our *rancho grande*. Duke and I waited for the trucks and cars to give way, then crossed the highway at a walk and took off at a run the moment we hit dirt again.

We were in marshlands just this side of the railroad tracks that went from San Diego all the way up to Los Angeles. I galloped Duke under the railroad bridge, hoping that no train came flying by overhead, scaring the living shit out of Duke and me. Then I saw them; the waves rolling up on the shores of Buccaneer Beach just ahead of us.

It was a beautiful low tide. The Father Sun was just beginning to set into the sea. The whole sky to the west was painted in colors of pink and orange and red, with streaks of lavender and silver. Duke was so happy to see the water that he wanted to gallop. I had to hold him back until we got to the sand. Once we were on the sand, which glistened like a mirror, he wanted to run out into the water itself. What the hell, I gave him his head and he turned north instead of south—the way I would've gone—and he was running wide open in the shallow surf like he knew where he was going. I wondered if he, too, was being guided by the Holy Hands of God, massaging him behind his ears.

Once again I remembered that this was what my brother Joseph had wanted to do—come riding to the beach on Midnight Duke.

"CHAVABOY!" I called out as we raced in the surf. "HERE WE ARE, Duke and I! Just like you wanted to do before you—"

I didn't get to finish my words. Suddenly Duke stopped, just like that, on a dime, and he turned his big long horse face towards the incoming waves, snapping his ears forward like he was listening to something. I took off my hat and turned my head, wanting to hear, too, but I couldn't make anything out, except the huge roar of the waves. But Duke seemed to really hear something, because he suddenly SCREECHED OUT, so loud and strong that his whole rib cage shook between my legs, startling the shit out of me.

I couldn't figure out what was going on. Who the hell was he calling to? I couldn't hear anything and I didn't see any other horses anywhere, up or down the beach. Then I saw something that looked like the top of a rock, a great big black rock, just this side of the huge incoming waves. But I couldn't figure out why a horse would be calling to a rock.

That was when I first saw the fins, and they were coming in with the waves, just to one side of the big rock, straight towards us. Duke was calling to them again and again with such power that I felt that his vibrating body was going to throw me right out of the saddle.

I got scared.

I couldn't swim very well. And were these sharks or what? And here they were, getting larger and larger, as they kept coming right towards us.

Then, I don't know why, but that little humming at the back of my head began to speak to me. Not just purr. But actually speak, telling me to keep calm, to remember that dogs, horses, cats, all animals could hear and smell and sense things far beyond all of my human senses, so I must just trust that Duke knew what he was doing, just as I accepted and trusted a dog to give the warning of an approaching stranger way before I, a human, even knew anything about the situation. I nodded, trusting the little voice that was speaking to me right along with the humming.

A great calmness came over me, as I watched these huge, thick-bodied, dark gray beasts rolling in on the waves. They were surfing into the shore right towards us. I now realized that they weren't sharks. No, they were dolphins, or porpoises. And there were two of them, and then I saw two more, and three more, and they were all now making eerie sounds to Duke as they kept swimming and rolling in with the waves.

Duke began pawing at the water with his right front hoof and continued screeching back to them. When they got in real close to us, he arched his neck and began to make a low guttural sound, and they began making chirping sounds right back to him.

The quick vibrations behind my ears spread to my whole head, and I began to understand things that I'd never understood before. My entire brain was now talking to me as it had never spoken to me before. And it was beautiful. The waves now had magic faces and I could see that each wave was alive. Everything, all around me, was now alive in a whole new way that I'd never seen. Duke was now humming, purring, too, in quick little vibrations, as he kept giving those low guttural sounds.

The dolphins were rolling their own sounds to Duke and me in the exact same, low gut-vibrations that I was feel-hearing between my ears.

I smiled. What did this mean, that the whole wide world could talk to one another through humming vibrations? This made a lot of blessed—yes, I mean, blessed horse sense.

The low guttural sounds now changed and became happy little high-pitched chirps. Hearing this, Duke got so wildly happy and excited that he suddenly lunged forward to go out into the sea and be with the dolphins. But I was scared of going any deeper, and the water was getting deeper and deeper with every lunge he took.

I jerked on the reins with all my strength, trying to stop Duke. But then he did something that I'd never seen a horse do in all my life. He turned his head sideways, grabbed hold of the side of the bit with his teeth, and jerked the reins completely out of my hands.

My God, Duke had just taken over his own reining! There was nothing I could do. I'd been a fool all these years to ever think that we people, who trained horses, were the ones in charge. I could now clearly see that horses could outmuscle us and outthink us, any time they wished. Duke was now his own boss.

I held on for dear life, praying as fast as I could. Duke was now swimming and the dolphins were all around us, and they were huge. I couldn't believe it, their bodies were as big as Duke's. And they just continued talking in low guttural sounds, then in real fast high-pitched screeching. Their conversation went on and on. They were just like *familia,* being so excited and happy to see each other. The last of the Father Sun was setting and long shadows were coming in all around us in wonderful colors of red and gold. The waves were now sweeping up over me and the saddle. I had my chest down by the saddle horn with my face in the horse's mane, trying my best to not get swept off by the seawater.

This was when I thought I saw something out of the corner of my eyes coming across the water. I lifted my head and saw that it was my brother Joseph, and he was skipping along the top of the water with another guy. They were just beyond the great big rock, where the ocean was calmer. The light of the fading day surrounded them and they looked like they were Angels having great fun as they made their way over the water. All at once I understood—oh, my Lord God—this other young man was Jesus Christ, Himself!

I quickly made the sign of the cross over myself, and the humming behind my left ear stopped, just like that, and instantly, I was sailing— knowing everything. Duke and the dolphins were cousins. Horses had originally come from the sea, as did all of life. And now they were simply saying hello because they were so happy to see each other after so many years.

Tears came to my eyes. I'd never seen Jesus before. In the past, I'd always only felt Him. My brother Joseph had gone up to Heaven and met Jesus, and now they'd both come back so that my brother could play in the surf with Duke and me.

I continued crying—I was so happy! My big brother had gotten his ride in the surf after all. It was dark by the time Duke and I started back for home. We went south on the beach and climbed the bluff alongside the steps at Cassidy Street, then headed for home.

I had no fear. The first Holy Star of the night had just come out, and the streets of South Oceanside were wide and clean and safe. I'd found my "place," yes, here, inside of me.

"Thank you, *Papito,*" I said. "*Gracias con todo mi corazón!* I guess that You really do know what You're doing. Thanks, *Amigo.*"

I swear that the Star above me started blinking at me. By the time I got to our big white gates, I just knew that I was now seeing with my Heart-Eyes, as my *mamagrande* had always told me that we humans needed to do before we could reenter the Sacred Garden *de Papito.*

CHAPTER nineteen

That night, I didn't tell my parents about the dolphins, but I did tell my little sister Linda, because I figured that she was still young enough to hear all this without getting scared. She loved it, and wanted to tell our parents.

"No, Linda," I said. "They might sell Duke."

"Why would they sell Duke?" she asked.

"Don't you remember how mad they got when we told them about Shep and they fired *Rosa* and *Emilio?*"

She began to cry. "Why can't we talk to our parents about these things? I want to tell them."

"We can't. They're too old."

"Well, then I never want to get old," she said.

And I agreed with her. I, too, was beginning to think that getting old wasn't good. Maybe this is why Joseph had left so young.

Then it was time for me to start school. I would be going to the Catholic school out at the San Luis Rey Mission, and this school, which was attached to the mission, was nicknamed the Academy of the Little Weed and it was mostly a girls' school, but they did allow boys to attend up to the fifth grade.

I was pretty apprehensive the first day my parents drove me to school, because ever since I'd been little, I'd been scared of nuns, with their dark clothes and long robes. The first time I saw a nun—I'd been about three years old—I'd yelled, *"COO-COO!"* My way of saying "ghost," and took off running in fear. I was older now, I was nine, and I'd had five years of public education to toughen me up, so I figured that I'd have a pretty good chance doing okay at this school. And I did for the first few weeks, but then one day, I blew it. This morning our nun was teaching us the Story of Creation when I raised up my hand and said that she had it all wrong.

"Oh, is that so?" she said to me. "All right, since you obviously know more than I do, why don't you come up to the front and teach the class."

In my ignorance, I didn't realize that she'd just said this to intimidate me and get me to shut the hell up, so I said, "Sure," and I got out of my chair and went to the front of the fourth grade class. Instantly, the humming began and that quiet little purring voice.

"You see," I said, "originally there were Two Sister Planets. Twin Earths, in fact, and when the great flood happened on our other *planeta*, thousands people got all the plants and animals off of our sister *planeta* and put them on a great sailing ship, which was almost as big as all of San Diego County, with hills and valleys and lakes."

The kids got all excited and some said that this made a lot more sense than how it was written in the Bible, with Noah's puny little ark.

"Sure, it does," I said to these kids, "because, you see, also, originally there were two Bibles, one for girls and one for boys so that men and women would know how to raise their kids, because you see, the real reason that we're here on Earth is to plant—"

But I never got to finish my words, because the nun, who'd sat down in the back to listen to me, now leaped up and came racing to the front of the classroom.

"That's enough! Stop this!" she yelled. "How did you dream up all these lies! There were never two Bibles!"

"I didn't dream up any of this," I said. "This was all told to me by my *mamagrande,* and also by my dad and mom."

"Are they authorities of the Bible?"

"Oh, yeah, sure, and gambling, too. Because, you see, all of life is a gamble, and so at gambling we must be king."

She suddenly looked very confused.

"Look, just let me go on," I said, "I was just getting to the good part. You see, we never lost the Garden of Eden. We just got fat and lazy and too self-important to keep planting the Holy Seeds that God sent us to plant here on Earth for His Garden."

"God sent us to plant the Garden of Eden?" asked the nun, turning very pale.

"Yes, exactly," I said, feeling good that she was finally catching on. "Just as He has been sending us to other Earths for millions and millions of years. We're all Walking Stars, see?"

In a flash, her mood changed and she lunged at me, gripped me by the ear, twisting so hard that I screamed in pain, and dragged me out of the classroom, yelling a word that I'd never heard before, "Blasphemy!"

I was taken to the head nun, a real old one, and the priest was called in and I heard them talk about my parents and my parents' money and it was decided that they wouldn't dismiss me, because of my parents being rich, but that I'd be kept away from all the other kids during recess from now on so I wouldn't contaminate their minds.

Then the next day after school two priests came to our home to see my parents. My mother made *yerba buena* tea and cut up some Mexican sweet bread into smaller pieces to serve the two men of God. They then explained to my parents what I'd said at school, and they added that they were sure that my parents had never told me any of this nonsense, being devout Catholics, so I must've gotten my wild ideas about twin sister planets from someone else.

I could see that my mother was all upset. She immediately told them how they'd had to fire *Rosa* and *Emilio,* two ignorant Indian people, because of all the old Mexican superstitions that they'd been trying to put into my sister's and my head.

"Just wait," I said, speaking up. "This isn't true. *Rosa* and *Emilio* never told us about the two sister *planetas*. It was you, *mama,* and your mother who told me the story of—"

But I got cut off before I could finish what I was saying, and told that I'd have to leave the room if I couldn't keep still. The two priests ate all the *pan dulce* and drank all the tea, decided to stay for dinner, had several drinks with my dad, and then they concluded their visit with telling my parents to also not speak any more Spanish to my little sister and me at home, that this was only hindering us from getting the best American education possible.

After the two priests left, it was really strange for my sister and me to hear our parents trying to speak to us in only English. My dad finally started laughing when we came to realize that Linda didn't know the difference between the two languages. She thought that a mixture of Spanish and English was a language all of its own, like she'd say, "Lets get *el caballo*-horse by his *pescuezo*-neck." Also, because I'd had such a difficult time learning how to read in public school, it was suggested by the two priests that, for a little extra money, I'd be kept away from the other students and given private tutoring by one of the young convent student nuns.

The young student nun that taught me privately was so kind and nice and beautiful that it was love at first sight. Within a week I was bringing her flowers almost every day, but then when I proposed marriage to her—because she was the smartest, strongest, and most beautiful woman I'd ever met—she told me that she was already married to Jesus.

"But He's dead, you know," I said.

"I'm married to Him in Spirit," she said.

"Oh, then that's okay," I said, laughing. "In Spirit, I've seen him, too. So you can then have two husbands like Mary—Joseph and God. I'll be your Earth husband, so we can kiss and have babies."

She smiled, and was just going to say something—I guess, maybe accept my proposal—when that real old nun rushed into the room, the one who'd called the priest, and slapped me so hard that I was knocked out of my chair. Then she began to hit the younger nun, knocking the habit off her head.

This was when I saw that my little nun had the most beautiful reddish-brown hair in all the world. It looked just like my horse Caroline's, the sorrel. I jumped up and attacked, biting the old nun on the back of her leg so hard that she SCREAMED out in pain, swinging around and around, trying to get me off her leg. She called me the

Devil, and sent for the priest again, and boy, did he get mad at me this time, when he found out which nun I'd proposed marriage to. I guessed he was in love with her, too.

I was now called an "infa-something," locked in a closet with brooms and mops, and every day I'd be marched to church to sit alone so I could repent for my terrible sins. But I could never figure out which terrible sins I should repent for, so I'd just sit there in the cool darkness of the church all alone, and I began to like it.

Sitting quietly one day I realized that all the statues and pictures on the walls of the church were talking to one another, just like Duke and those dolphins had spoken to one another. I liked this. It almost sounded like the whole church was alive with a symphony of purring. Soon I began to like getting punished, just so I could be sent to the church to be alone. I began to draw stars once again. Holy Stars. And now, since I'd seen my brother with Jesus down on the beach, I knew why I loved to draw stars.

Drawing stars was what helped us kids keep our soul-memories alive. Drawing stars was what kept us from going crazy, surrounded by all the doubt and fear that adults had. I began to notice that I wasn't the only one drawing stars. So were half of the kids in my school. I'd see stars in their workbooks. I'd see stars on the cover of their binders. I'd see stars on the palms of their hands. But of this I told no one, except my sister Linda. I was in enough trouble already.

One day I was walking to the church alone to repent for my sins when I spotted my little nun. "Hello!" I yelled to her. I hadn't seen her since the day that I'd proposed to her. But when she saw me, her face filled with terror, real terror, and she turned and ran away from me as fast as she could.

I ran after her, across the grass, and into the building she'd entered. Here there were nuns—dressed just like her—everywhere. I was stopped by one nun and asked what I was doing in their private quarters. I tried to explain, but I was grabbed and slapped before I could say anything. Then one nun—I'll never forget—said that I was the Devil boy who'd caused Teresa all her trouble. Two big, strong nuns grabbed me real hard and took me out of the building, back across the grass, into another building, and I was locked in a dark, smelly room where I could cause no more problems. Then came the priest, the same one who'd gotten so angry when I'd proposed to the little nun. He screamed and shook me so much that day that I puked.

When I got home after school, I immediately saddled up Caroline and went running along the railroad tracks. I climbed up the hill from the tracks to the green grass of the cemetery and tied Caroline to the fence that ran along the outside of the burial grounds. I climbed the fence and ran up to the big white cross with Jesus and Mary.

"What the hell is going on?" I yelled at Jesus, I was so mad. "Eh, what's wrong with You and my brother? I thought that You two were looking out for me! I didn't do anything wrong, and yet I just keep getting in trouble. THAT'S NOT FAIR, DAMNIT! And yes, I mean DAMNIT! DAMNIT! DAMNIT!"

But Jesus and my brother wouldn't talk to me. I guess that They didn't like me cursing at them. It was Mary who finally spoke.

"Come to me," she said in a lovely soft voice as she put out her arms for me, "and let me hold you. We all know how hard it is for you, and We love you with all Our Hearts."

I went to Mary. She wore such beautiful, colorful clothes, and I lay down on the grass beside her as she held me. It felt so good to be in her arms. I relaxed and cried and cried, and soon the rage began to leave my body. Then I felt two hands begin to massage my back. It was *Chavaboy*, I just knew it.

"Joseph," I said to him without turning to look at him, "really, it should've been me to go, not you. I'm not smart enough to figure out what to do here," I said with tears coming to my eyes. "I'm a *burro*, the dumbest kid in all of our school. You should've stayed and become a lawyer like you said you were going to do. What can I become, except maybe just stupider? Damnit, Joseph," I added, suddenly getting angry again. "WHY DID YOU LEAVE ME! You never had any problems learning how to read. You were a GENIUS! I should've died! NOT YOU!"

The massaging of my shoulders stopped just like that, and I realized that I shouldn't have gotten mad and used the word "damn," but I was so damn, damn, damn, DAMN MAD that it was hard not to say "damn" when I felt like the whole world was against me.

I must've fallen asleep, because the next thing I knew, I awoke feeling really good. Wonderful, in fact. I stretched and yawned and felt like I'd maybe gone up to Heaven to visit for a little while. I looked to the west and saw the tall, flat *mesa* where Shep had leaped into the sky to intercept my brother's soul, and I just knew that this was, indeed, a Sacred, Holy Place where I was. I decided to start coming to this place, our "place," every day after school. This I would never tell anyone, except my sister Linda.

The following year, our little sister Teresita was born, and our father took Linda and me to see her at the hospital, but they wouldn't let us in because we were too young. Our father winked at us, then he took us around to the back of the hospital. He opened a window and lifted Linda and me up to the window so we could crawl in and meet our baby sister. This was so exciting! We were all a little *familia* hiding from the hospital authorities.

This was the same year that my parents sent me to a new Catholic school, Saint Mary's Star of the Sea at Hill Street and Wisconsin in Oceanside. At this new school—it had just opened—things got much better for me, and I think that it was partly because of the school's name, "Star of the Sea." But still, I could see that I really wasn't learning how to read any better. It was just that now we were no longer required to stand up and read aloud in class, so no one realized that I couldn't read.

This was also when I met Nick, Clare, Sally, and Dave Rorick, whose father owned Rorick Buick in downtown Oceanside. I became good friends with Nick who was in my same grade and his friend Dennis Tico, who was the best-looking guy in all of our school, or so all the girls said. Myself, I'd started to put on weight ever since my brother's death, and then gained even more weight when I hadn't been allowed to play with the other kids at the San Luis Rey Mission school.

I went to Saint Mary's Star of the Sea for two years, and even though I never learned to read any better, I was able to at least hide this fact from the other kids. For the first time since I started school, I wasn't called stupid or slow. I got to play on the playground with the other kids and everything was going pretty well . . . until one day when a real poor-looking blondish kid was put in our class. He had three younger sisters and they all wore dirty, old-looking clothes and they'd pick their noses without caring if anyone was watching.

His name was Augustus and he was in the same class as Nick's and mine. They had us in alphabetical order by our last names, and Augustus was placed right in front of me. Just before lunchtime, when it was time for us to stand up, put our hands together and pray, it was easy to see that Augustus really didn't want to do it.

"Put your hands together and pray with us," said our teacher, Sister Michael Mary, who was also the principal.

Augustus just shook his head, looking all beaten down.

"I said," said Sister Michael Mary, raising her voice, "put your two hands together this moment, and do your prayers with the rest of us!"

Augustus was hurting, I could tell. So I spoke up for him. "But sister, you just gave us a long talk on free will, so if he doesn't want to pray, then he doesn't have to."

But my words didn't help. Sister Michael Mary, who was usually pretty calm and fair-minded, now rushed down the aisle and grabbed Augustus by the shoulders and shook him real hard. I guessed that she thought that it was he who'd spoken up.

"Don't talk to me about free will!" she yelled. "We took your family in when we had no more room, and you will now pray like everyone else when it's prayer time!"

"But sister," I said, still trying to help Augustus out, "this goes against everything that you've been teaching us all month. You said that free will gives us the freedom of choice, so he has the God-given right to—"

"You haven't been here a month!" she screamed at Augustus, grabbing his hands and slapping them together. "Now pray! You hear me, PRAY!"

She was so angry that she couldn't see that it wasn't Augustus who was talking. It was me who was behind him. She turned and went back up to the head of the classroom and continued leading us in prayer. Augustus just dropped his hands once again, and I could see from the back of his shoulders that someone, somewhere had "broke" him real bad like so many stupid, mean cowboys did with horses. But Sister Michael Mary couldn't see any of this, and when she saw him drop his hands again, she came flying across the classroom. She slapped him in the face until Augustus finally put his hands together to pray.

The moment we went out of the classroom for lunch, Augustus took off after me, yelling that he was going to kill me for getting him in trouble.

I took off running, but he caught up with me at the end of the playground. "I'm sorry," I said, between heaves for air. "I'm really sorry! I was just trying to help you."

"Well, don't try and help people, you fool!" he said, hitting me two or three more times. I didn't fight back. I could see that he was just upset. "I never told them to take us in," he said. "Our pa left us, going back to Texas, and my ma didn't know what else to do to feed us except take us to the priest, like her ma always done."

He quit hitting me and started crying. I apologized some more. He finally forgave me, saying that he knew that I'd meant good. We then walked back to the school buildings where everyone was having lunch. I got my lunch box, and when I saw that Augustus had no food, I split my lunch with him, and he told me the strangest thing—that his grama was Mexican and she, too, had always made egg and *chorizo burritos* for him. I couldn't believe it, he was blondish, and yet he was part *Mexicano*.

The next day, I brought extra *burritos* for Augustus and his two little sisters, but they were gone. They never came back to school. I guess that they'd gone back to Texas to try and find their dad.

About a week later, our new, young priest from the Oceanside parish came to our school. His arms were thick and real hairy and he informed us that a great opportunity was available for us students, if we said one rosary every day for world peace for two straight weeks. If we accomplished this, we'd get a special picture of a saint to carry in our pocket, and we would also have a guaranteed passage into Heaven when we died.

I wanted to ask the priest how this was possible, because it didn't make any good horse sense to me. Did this mean that after we said these rosaries for two weeks, we could go out and do bad, mean things and still have a guaranteed passage to Heaven? But by now, I'd been beaten enough and seen enough slapping and yelling any time one of us kids asked a good question, that I knew better than to ask this.

My God, back in third grade, when I'd told What-A-King that I'd figured out that there was no Santa Claus, his parents had told him that it was true, there was no Santa, it was they, our parents, who gave us our Christmas presents. They also told him to keep away from me, because any kid who went around telling other kids such things was a bad person. So then, I guessed a kid was bad if he questioned, and was a no-good rat if he figured things out. My dad had sure been right when he'd told me that in every game there were two sets of rules: one for the public, and another for the insiders to use only for themselves.

Realizing this, I began to feel that purring behind my left ear. Then I heard the "Ghost Riders" song that my brother had listened to over and over again the last few months of his life. I smiled. So this meant that Joseph was now up in Heaven with *Papito,* helping Him to massage the back of my head.

I decided to say nothing to the priest or nun of what I was thinking, and sign up for these rosaries, too. Why not? Anything that would give me a better chance of getting into Heaven so I could be with my brother when I died made a lot of good, common horse sense to me.

But then, to my complete surprise, when I raised my hand like the other kids and said that I'd also like to sign up for these two weeks of saying a rosary a day, Sister Michael Mary looked at me and said, "Are you sure you want to do this?"

"Yes," I said.

"Do you understand that this is a serious commitment?"

"Yes," I said again.

"Okay," she said, "but I'll be checking on you every day to make sure that you did say your rosary."

My heart started pounding. "Does this mean that you'll be checking on the other kids, too?" I asked.

She exploded! "You will not QUESTION MY AUTHORITY! I will check on whom and what I see appropriate!"

"Well, what is appropriate? Because you only asked me if I understood that this was a serious whatever-you-said, and you didn't ask any of the other kids when they raised their hands."

And I knew that I was a *burro* to have said all this. I should've just kept quiet. And I was right, because she now came flying across the classroom with her white robe and black habit flapping like wings, scaring the pee out of me. She got right in my face, and I could see in her eyes that she wanted to grab me and strangle me to death, but thank God, she didn't. Instead, she caught her breath, heaving and puffing, and said, "You'll be getting your walking papers if you keep this up!"

"What are those?" I asked, not understanding. "I've never seen papers that got legs and can walk?"

Some kids started giggling.

"SILENCE!" she screamed, whirling on everyone. The kids froze, and she turned back to me. "Don't you think that we don't know about your past," she said to me. "We were warned about you by the Mission!"

And saying this, she stared at me, giving me that mean eye that only the best of the old nuns can give. I said nothing more, and that afternoon after school, I asked Nick, who was the smartest kid in the whole school, what he thought.

"About what?" he asked.

We were on our bikes, on our way to his house, which was really two houses right next to each other by the Oceanside pier. They were brown and huge and each was two stories.

"About how Sister Michael Mary didn't want to answer my question about what's appropriate, and she went crazy."

"You had it coming," he said.

"I had it coming?" I said. "But Nick, it wasn't fair that she was only going to check on me and not the other kids."

"Maybe. Or maybe not," he said. "You're not one of her most active participants in extra curriculum, so it seemed strange—for me, too—that you'd volunteer. I'm not going to volunteer."

I hadn't really understood everything that he'd said, but still I wanted to tell him that this wasn't what was going on. In my opinion, I thought that she'd singled me out because I was—but I couldn't bring myself to use the word "Mexican." Because my new friends at this school didn't really see me this way. I said nothing more.

We parked our bikes in the backyard of the two huge brown houses. The backyard had green lawns, beautiful flowers, red brick walkways, tall trees, a huge fish pond, and white furniture. I visited with him and his sister Clare. Then we went uptown to take Clare to her piano lesson. Nick went back home so he could go to the beach with Dennis Tico. I went to our local stationery store to buy another fountain pen. A few weeks ago, I'd discovered peacock-blue ink. It was love at first sight. This was the exact shade of blue that I'd been looking for all my life, so I could color-in my Stars, then jump in them and be gone.

For me, blue wasn't just blue anymore. I was beginning to see that there were at least ten different shades of blue. That's what I had seen when I'd been on Duke, swimming with the dolphins. I'd seen all the water alive. Nothing was just what it was. Everything had many shades of reality. Blue wasn't just blue. Water wasn't just water. No, everything was alive with all these different shades of color and light. And every shade of color or light felt different so deep inside of me. Like I'd found that when I colored-in my star with this exact shade of peacock blue, then added that tiny touch of red and yellow, immediately the purring began behind my left ear. Then the whole world suddenly became alive all around me, just like when I'd seen Jesus and my brother. Color and light, these were the "eyes" of my purring. The eyes of *Papito* seeing through us humans.

I parked my bike outside of the stationery store. It was a new bike, just like Nick's, with slender tires. I hardly rode my big old heavy Schwinn, anymore. I went into the stationery store all excited, with my pen in hand that I'd purchased the week before. I wasn't able to find what I wanted, so I walked up to the cashier.

"Ma'dam," I said really happily, "I'd like to buy another pen like this one and another jar of your peacock-blue ink."

Instead of addressing my question, the woman reached over the counter and snatched the pen that I was holding right out of my hand.

"Give me back that pen!" she snapped. "You can't just take our merchandise!"

"But ma'dam," I said, suddenly getting all confused inside, "I didn't take it. I bought it last week."

"How can I know that?" she said. "You're a Mexican, and everyone knows that Mexicans are thieves and can't be trusted! Now out of here!"

I was shaking so hard by the time I went out the door that I couldn't even get on my bike. She'd looked at me with that exact same kind of look that Sister Michael Mary had looked at me. And my friend Nick hadn't seen it. But how could he? He'd spoken English on his first day of school and he'd learned how to read immediately. Hell, last week when we'd been tested for reading, he'd tested six grade levels ahead of everyone, reading at an advanced high school level. And me, I'd made sure that I was home sick on the day we were tested for reading, because I knew that I was still below a third-grade reading level.

I took a deep breath. So this was it? As long as people saw me as a Mexican, I could never be trusted by anyone, not even by a nun who was supposed to be so wise and close to God. No wonder all those quicker-learning Mexicans had said they were Spanish or French—anything but *Mexicano*.

Realizing this, I got to feeling so bad and hopeless that I wanted to tear the brown skin off my body. I was shaking so bad by the time I got up to the street where Nick's sister, Clare, was having her piano lesson, that I didn't want to see her.

I was supposed to accompany her home, but I felt so heartbroken and angry that I couldn't see anyone. When I saw Clare come out of the house where she'd taken her lesson, I felt so ashamed that she might see me crying that I took off on my bike without saying a word to her.

CHAPTER twenty

I was in the seventh grade when I started going to the Army and Navy
Academy. My little sister Teresita was talking pretty well by now. It
was early evening and school wouldn't really start until the next day, but
this evening we, the new day students, had to go for a two-hour orientation. All the way home, my little sister Teresita kept laughing and moving her arms up and down, pretending like she was marching, then
she'd bark orders at me just as she'd seen my fellow cadet leaders bark
at me. I didn't think it was funny in the least, but Teresita and Linda
and my parents thought that it was hilarious.

The following day, on my first official day of school at the Army Navy
Academy, my *problemas* immediately got started with Moses. He was a
captain, as were all of our teachers in the lower school, meaning the
seventh and eighth grades. It was late afternoon and we were done with
our classroom studies and our marching. Now we changed clothes, got
out of our uniforms, and went to do sports. I wasn't very tall, but I was
one of the bigger kids, especially since I'd put on all that weight since
Joseph's death. I'd never played baseball or football, so I had no idea
what was going on when they threw the football at me. I ducked, so it
wouldn't hit me.

Moses got mad and told me that my brother Joseph had been one
hell of a football player, so for me to wake up, catch the ball, and run
with it through the line of scrimmage. I had no idea what this word
"scrimmage" meant and I didn't see any line to run through, either. But
I could see that Moses was getting madder and madder and he finally
called the three biggest guys aside, Wallrick, Altomar, a Mexican guy,
and this other huge cadet named Williams. These three guys had little

smiles on their faces, and they hit me really hard the next time I was given the ball. I wanted to cry. My brother had gotten sick and died because of a football injury. I didn't want to get sick and die. This was when a great big heavyset cadet named Hillam came to me, helped me up, and spoke to me in a kind voice.

"Do you know why they want you to run with the ball?"

"No," I said.

He laughed. "It's because that's how we score."

"Score what?"

"Points."

"Points for what?"

"To win."

"Oh, like my cousin Chemo used to do for the football team in Oceanside. But he didn't run at any line. He'd run around people, across the field to those tall posts."

"Yes, but in practice, we don't run all the way to those posts. To save time, we run at the line of scrimmage, which in football jargon means the line of the opposing team."

"Oh," I said. "Well, then, why doesn't Moses just say that?"

"Because he likes to show off how much football jargon he knows," said Hillam, laughing.

This guy Hillam, I immediately liked. He made good, plain horse sense. It was then time for us to go in, and all the way back to our barracks, the other cadets kept pushing and shoving each other, then they'd hit me when I wouldn't push and shove back.

We had to strip to shower. I'd never stripped in public. The guys were laughing and snapping at each other with their towels, and when we got in the showers, Altomar, a guy way bigger than me, looked at my penis and said, "You call that a dick! Hell, you don't even got any hair!" He started laughing. Then under his breath he said to me, "Your dad might have the biggest house in all the area and think he's a big shot with his big cigars and Cadillac, but here you're nothing but dead meat, punk!"

And saying this, he hit me in the stomach so hard that I went down, gasping for air. Altomar laughed and said, "What happened? Did you slip?" And he grabbed me and jerked me to my feet, and continued showering as if nothing had happened.

That was when Wallrick, our class leader, came in, and he and Altomar started talking like they were best friends. And it turned out that they were. Both already had hair all around their private parts, and they were from Vista, and Vista now had its own high school, so they were rivals of Oceanside.

I was stunned. I had no idea how Altomar even knew about our home. My parents' Cadillac, of course, everyone had seen when they brought me to school yesterday evening for orientation. I never showered that day. No, instead, I got dressed and went outside to wait for the bus to take us day students home. For the next five years at the Army Navy Academy, I never once showered at school. I'd always wait until I got home. This was, also, when I . . . began to have trouble peeing in public toilets, and then finally even at home in private.

By the end of the first week at the Academy, I understood very clearly that Moses was best friends with Wallrick, Altomar, and a couple of the other bullies. In fact, I began to see that it was he who was instilling in these guys the idea that they had to be real men, and real men were tough, gave no quarter—which I thought was really funny. Did this mean that they gave half-dollars?

The academy became my new living hell, just as the Little Weed had been for me out in the San Luis Rey Valley and the Oceanside Public School System before that. They all saw their job was to "break" us, not to befriend and *"amanzar"* us. They hadn't been raised like women for the seven years of their life, so they had no idea how to be gentle, patient cowboys, Christians, or cadets.

Many a night I cried myself to sleep, asking God what was going on, but I could no longer hear that little voice deep inside of me nor feel that humming behind my left ear, even when I drew stars and colored them. I felt abandoned by *Papito!* I felt abandoned by my brother Joseph, Shep, Sam, and my two grandmothers, and even Jesus and Mary.

Then came that short, muscular, blond substitute teacher and he put his poster on the wall of skiing in Colorado, and hope was born! For three glorious days I once more felt the humming and I could hear God's purring Voice of Love, too. But when the substitute teacher left, I began to hate that he'd ever come into my life and given me even a little glimpse of light in my awful darkness.

I could now see that the nuns had been right when they'd called me the Devil, because now that Moses had given me that *F,* without even reading my paper, then ridiculed the name that my dad had given me, I was thinking evil thoughts every time I'd see him. I was going to kill him. I was going to do as evil does.

This was when I called up information to get Captain Moses's phone number and address. Getting it, I dialed his number. He answered. I heard him say, "Hello, hello," over and over again, and then I hung up. Now I was the one who was smiling.

I decided to call information again and get that playground teacher's phone number and address, the one who'd tortured us in kindergarten. I got it.

I called her. She answered, saying, "Hello, hello," too.

I hung up, and I just couldn't stop smiling. Now I had a plan. Now I had a reason to live. Now I was going to kill these two teachers and that priest and nun who'd abused me. I was going to blow them all into smithereens! And I'd do it all in one night!

But then I remembered that I didn't have a permit to drive a car yet, so I'd have a hard time getting everyone in one night on foot or on my bike. It probably made a lot more horse sense to only dynamite or kill one person per night.

I had a mission. It felt wonderful!

Late one night after school, I rode my bike over to the Army Navy Academy. I'd learned that most of our school faculty lived within a few blocks of our campus. Having written Moses's address and phone number on a piece of paper, I was able to find his home. I rode past his house several times. On my last pass, I saw a woman through the kitchen window and I realized that she was our school librarian. I liked her. She was a nice person. I couldn't figure out why she would be at Moses's house. I pedaled home.

That week, I asked some questions at school and found out that Captain Moses was married to our librarian, and that they had a little girl. This pissed me off! Now I couldn't just blow up Moses's home, because I didn't want to hurt our librarian and their child. I'd have to shoot Moses. Yes, I'd use my dad's old 30/30 Winchester and shoot him through their kitchen window. But I got to figuring and decided that a 30/30 was too big and noisy. Hell, I'd just use my .22 pump action Winchester. After all, we killed our thousand pound steers with a .22

My secret life was so exciting! I was now shooting well over five hundred rounds of ammunition per week. I was getting ten or twelve rabbits every time I went hunting. All that training that I received from Shep on how to stalk game was now really paying off.

One night, I decided to bike past the playground teacher's house. But I didn't take the piece of paper on which I'd written down her phone number and address. No, after that first night that I'd biked past Moses's house, I destroyed the piece of paper on which I'd written down his phone number and address. Hell, if a cop had stopped me and asked me what I was doing, and then searched me and found that piece of paper, my goose would've been cooked. Because once they'd begin questioning me, I was sure that I would've broke down and started crying and told them everything. Over and over my dad had explained to me that a man had to keep his cards close to his chest, have a plan, think ahead, and always bring into consideration all the things that could go wrong. That a chain was only as strong as its weakest link. If I planned to kill all these teachers, and that priest and nun who'd been cruel to me, I had to be the strongest, best thought-out hunter-stalker I could be.

Her house was easy to find. It was just off Wisconsin, two streets east of the railroad tracks. On my last pass by her house that night, I spotted her. She was coming up from the beach with her dog. Her dog immediately reacted to me. I guess that he could smell the hate I had for her, because he started not just barking, but going crazy. She never recognized me, and this was good. I decided that I wouldn't come back by her place until I was ready to do her in. And about her dog, what the hell, I'd just send him to damnation, too, when I blew up her house.

The school year was almost over. Soon it would be summertime. I decided not to do all of my killings this year. What the hell, if I got caught, it could ruin my summer. And this summer I'd be getting that .22 Smith & Wesson revolver with the six-inch barrel on a .38 frame that my dad had promised me, so I didn't want to miss my chance of getting this handgun and getting even better with weapons.

Also, I was now preoccupied with a few other things, like this one great big older guy who rode on the bus with us every day. His name was Workson. He lived in North Oceanside. He was a day student like me, a sophomore in high school, a basketball player, well over six feet, and he wore a ring with a wolf's head with tiny red eyes. Every day when we'd pick him up in the bus that took all of us day students to school, he'd start off his day by turning his ring around on his finger, and walk down the aisle, hitting all of us younger kids on the head, telling us that he couldn't wait until we'd gotten out of grammar school and were in high school, so that he could really give us our proper initiation, just as it had been done to him.

One day he'd hit me so hard that I'd started crying. He'd called me a little sissy girl and hit me six or seven more times, telling me that wolves came in packs, just like a platoon, and that I had to stop being a crybaby and grow up if I wanted to make it in high school at the Army Navy Academy.

He didn't know it, but that day, he made it to my kill list. After that, his daily blows to my head, which he called knockers, didn't hurt as much, even though many times my scalp would be bleeding by the time we'd get to school. The little ears on the wolf's head were pointed and would cut into my head like sharp little spikes. Our bus driver saw this every day, but he only laughed along with all the other older cadets, saying, "Boys will be boys," just as Hans had done that night I'd had my encounter with the huge, monstrous frog.

When that first year at the Academy was over, my dad took me down to Johnson's Sporting Goods store in downtown Oceanside and Mr. Johnson immediately brought out the .22 Smith & Wesson that my dad had promised to buy me. My God, the feel, the balance of this handgun was magic! This was no Saturday Night Special. This was a finely made weapon of utmost precision. Instantly, I became a great shot. Then I got a holster and ammo belt just like in the good old Wild West and I'd practice quick drawing and shooting.

By the end of that long hot summer, I was lightning fast and accurate, too. I decided that I wanted to kill Moses . . . face-to-face, like in the Old West at high noon on main street Carlsbad. I wasn't just going to blow his house to smithereens. No, I wanted him to see my eyes as I came walking down main street, and I'd explain to him why he was going to die.

"You're a bad teacher," I'd say, farting. "A cruel teacher, who likes teaching meanness to kids. This is why you are going to die."

I'd fast draw and gut-shoot him first, then I'd shoot each leg, and keep him alive so he could suffer and really understand why he was dying—I drew and shot well over five thousand rounds that long summer, pretending that I was killing Moses with every wonderful, well-placed round!

No longer did I hear the humming or purring behind my ears. But who cared? Now I heard the ringing of gunfire in my ears! And this felt ALMIGHTY GOOD!

It was my second year at the Academy, and a bunch of my fellow eighth-graders were weight-lifting behind a cottage on the grass. Only a few could standing press—without jerking—a hundred pounds over their head. Wallrick, Altomar, and a few others did one hundred and twenty over their head. These guys were really strong. They told me to give it a try. I couldn't even budge the hundred. They all laughed, and said that I had better work out real hard this year, because things were really going to get tough for all of us once we got into high school with the big guys. These words hit me like a ton of bricks. They were right. I could see it so clearly. It wasn't enough that I was good with guns. I had to get physically strong like I'd been in third grade, before I hadn't been allowed to play at recess with the other kids. I began to put more effort into the push-ups and other exercises they had us do.

A new teacher came to school. He was tall and had a loose hip gait like a Tennessee Walker horse when he walked. He kind of reminded me of the substitute teacher that I'd had the year before. This teacher's name was Brookheart. It was a beautiful name, I thought; a heart alongside of a brook. At the end of the year, he gave me a B by accident. I went to him and told him that I needed to talk to him privately. He said to meet him after school. I did.

"Sir," I said. He was a captain like Moses. "You made a mistake on my grade."

"You think so?"

"Yes," I said, "I checked all my past homework real carefully and I was supposed to get a D not a B."

"You mean, you're complaining because I gave you a B instead of a D."

I closed my eyes in concentration. "No, sir . . . I'm . . . I'm not complaining. What I'm trying to do is get this straight so you won't get in trouble."

He stared at me. "You're concerned about my welfare?"

I didn't know what the word "welfare" meant, so I shrugged. "I don't know about that," I said. "All I know is that I like you, respect you, and so—" I almost started crying, but didn't. And for the first time in years, I felt that humming start behind my left ear, then I felt it shoot across the back of my head to my right ear. Instantly, I just knew that *Papito Dios* was here with me once again, massaging me.

"Look," he said, "I'm not going to get in trouble. No one checks my grading, especially when they're Bs or As."

"Why not?" I said, before I realized that I'd even spoken.

Quickly, I cringed and ducked my head in case he took a swing at me. If there was one thing that I'd learned real well in school so far was not to question without being ready to get slapped or yelled at.

But he didn't take a swing at me. Instead, he sat back on the front of his desk, just as Mr. Swift had done for those three days, and he looked at me with a kindness that I rarely saw in teachers, but I did see more in dogs, cats, horses, goats, and sometimes even in cows and pigs, too. "Because As and Bs are what the administration and the parents want," he said.

"But don't they want to know if it's true?"

He laughed. "No, not really." Then out of the blue, he said, "Look, I know that Moses has something going with you, but I want you to realize that it really has nothing to do with you. You see, he never got above a corporal in the Regular Army—just like Hitler—and so they both have that same kind of small man's mentality."

I stared at him in utter shock. I didn't know that anyone knew about Moses torturing me, every chance he got.

"You don't really understand what I'm saying do you?"

"No, not quite, sir."

"Well, that's okay. Just remember that you probably don't even belong in this school. This school is populated mostly with spoiled rich kids who weren't handled very well at home. You have a real home, from what I've heard. The B remains, and don't take Moses too seriously. The man is pitiful."

I didn't know what this word "pitiful" meant, either, but it felt so good that I was going to be able to keep the B and he wasn't going to get in trouble. And yet, I hadn't earned the grade.

"Yes, what else is it?"

"Sir, I didn't earn that B."

"Oh, yes, you did!" he quickly said. "You earned it by having the honor of coming in and telling the truth, and that's rare! All right, enough! If you stay any longer, I'll make it an A."

I left as quickly as I could, feeling ten feet tall. Later, just before school was let out for the summer, I found that Captain Brookheart had actually been a major, or maybe even a colonel, in the real Army, but they'd asked him to go down in rank so it wouldn't embarrass all the rest of the school staff, especially the Colonel, who ran the whole school, but had never even been in the U.S. Army at all.

I came to realize that most of our faculty had fake rank, fake titles.

My whole world went into a spin. I wondered if this was also true of the nuns and priests and teachers in public school. Were they really trained, qualified teachers who knew what they were doing? Maybe they weren't, and this was why they'd been so mean and ignorant about the whole educational system. The truth was that they really didn't even want to educate us. They wanted to train us. That was why they felt that they needed to "break" us down first, get rid of our spirit, our guts, our soul, so that they could then remake us in their own image of being mean, scared, and all mixed up. Boy, was I ever glad that I'd been too stupid to learn how to read. This was probably why I hadn't been able to get "trained" into all of their *caca*.

No, I wasn't very smart, this I knew, but I was beginning to think that maybe, just maybe, I was some kind of crazy-*loco* genius, *burro* genius. I mean, to have been able to hold on to my Spirit for this long had to mean something.

CHAPTER **twenty-one**

That summer, I drove out with my dad to look at a great big ranch in the San Luis Rey Valley. Part of the ranch ran alongside the little airport that was located just south of the main riverbed. It was a beautiful place, full of hills and valleys and wild deer and quail and skunk and bobcat and even mountain lion. Its northern border butted up against Camp Pendleton for about a mile. My father asked me what I thought about it. I told him that I loved it. He said he loved it, too. In fact, it reminded him a little of his home back in *Los Altos de Jalisco*. So he purchased the huge spread, and I learned how to drive a Caterpillar tractor, plant hundreds of acres of hay, and then bail it. I worked from sunup to sundown with our workers from Mexico. I worked so hard that all night long, I dreamed of bailing hay, loading trucks, then unloading them into our barns. I didn't have a driver's license, but still I was driving trucks and tractors. This was the summer that I met John Folting, Terry Watson, and Ted Bourland, Little Richard, Bill Coe, and Eddie O'Neil. I didn't see as much of Nick Rorick anymore. All these other guys lived in South Oceanside. They became my first real group of friends, next to Nick Rorick and Jimmy Tucker and Jimmy's cousin, Michael.

On my first day back on the school bus to the Academy, I quickly got out of my seat the moment Workson, the guy who'd been hitting me on the head for two years with his wolf ring, got on the bus.

"I'm ready!" I said to him.

"Ready for what?" he said, looking surprised.

He was still a whole head taller than me, but I didn't give a shit. This was it. "You've been telling me for two years how you're going to really get me once I get into high school, so here I am. I'm in high school now. So come on! I'm ready! TRY AND GET ME!"

My heart was pounding a million miles an hour, but I knew what I was going to do. I had a plan. I was short, he was tall, so I was going to belt him in the balls, again and again till he bent over, then I'd grab him by the hair, jerk him down to me, and bite him on his long nose, tearing it off his face like I'd seen an old-timer bite the balls off of our goats and sheep.

But what Workson did next took me by complete surprise. He started laughing. This I hadn't been prepared for.

"Get away from me," he said, backing away. "You're crazy!"

My heart was beat, beat, BEATING!

I SCREAMED! I BELLOWED! I started kicking at the seats in our bus. Our bus driver pulled over. But this time our driver wasn't laughing and saying that boys will be boys. He wanted to know what was going on. I couldn't talk. I couldn't explain. Finally, Workson said that nothing was going on, that I was just a mixed up little freshman who didn't know how to take a joke.

My rage exploded! "IT WAS NO JOKE for my head to get BLOODY every morning for two years!" I screamed.

The driver told me to calm down and get hold of myself or he'd report me.

"REPORT ME!" I yelled. "This is all ass backwards! Where was your concern all these years when he'd hit us little kids! I'LL REPORT YOUR ASS, TOO!" I screamed.

"Okay, there will be no reporting," said our driver. He was scared. I could see it in his eyes. It had worked. My surprise attack had paid off. "Please, just take a seat. You've made your point."

I took a seat, sitting down, and tears of rage—not fear—began streaming down my face. I had my .22 Smith & Wesson revolver with me in my book bag. I would've belted him, bit him, shot him, and killed him! But he'd laughed and backed away, chickening-out on me, and so he'd slipped out of my death trap!

But at least I was now ready for Moses. Today, today, TODAY, I was going to KILL HIM! This was why I'd brought my revolver: to kill Moses in public. Face-to-face.

Then a strange thing happened. That day after classes, a bunch of cadets started lifting weights out on the grass behind a cottage once again. And these were the big high school cadets. Really big guys. Still, most of these guys couldn't press a full one hundred pounds over their heads without dipping their knees a little bit, then throwing the weight upward. But a lot of guys could. They asked me to try. I said no, because I knew I couldn't even budge a hundred pounds.

But they kept insisting, so I finally agreed, and walked up to the weights, took a few breaths, bent over to grab the bar, expecting it to be so heavy that I couldn't lift it off the ground at all, but, to my utter shock, the bar and weights went flying up and over my head with ease.

They added more weight and it ended up that I could do one hundred thirty pounds in an overhead press without dipping my knees at all. My God, I was only a freshman, and already I was one of the strongest guys in high school!

One cadet told me that maybe I should try out for the wrestling team. I'd never even known that our school had a wrestling team. But I'd always loved wrestling, so I immediately went to sign up.

At the office, I was told by a senior, who was already on the wrestling team, that they were all filled up. I didn't understand. He explained to me that they wrestled by weights, and that each weight class was already taken by an experienced wrestler, so I wouldn't have a chance of making the team. I said I understood, but still I wanted to work out anyway, that it really didn't matter to me if I made the team or not. Still he didn't want me, and said that maybe I was strong, as he'd heard I was, but I was too heavy to move with any kind of speed. I wondered if he didn't want me on the team because I was Mexican or did he really think this? Hell, he didn't know me. I'd been wrestling calves and goats and pigs all my life. I was lightning-fast.

"I still want to work out with you guys," I said.

"Okay," he said, "we could use extra meat to practice with."

My heart exploded! He hadn't been very smart to say this to me.

Now, I was determined to bust his ass. "Come on, give it to me, God!" I could hear my dad saying that his mother used to say when the going got tough. "This is our chance to move MOUNTAINS!"

Within a week, I beat that guy who'd called me "extra meat" and had been on the wrestling team for a couple of years. I started running and running, lost fifteen pounds, made the varsity team—not the junior varsity—and won nine out of twelve matches against juniors and seniors, and I was a freshman.

I forgot all about killing Moses with all this energy and time that I was putting out towards wrestling. I LOVED WRESTLING! And I also made some new friends: Juan Limberopulos from Guadalajara, Mexico; Mick McLeans from North Hollywood; Hawkins, our team captain, from Reno; and Fred Gunther from Temecula, just east of us.

Our coaches were two Marines from Camp Pendleton, and these guys were fantastic! One almost made it to the Olympic Games. Our sense of balance, we were taught, came first, then came speed, skill, endurance, and strength was last. I worked and worked with all my heart and soul, just as I'd done in the third grade with marbles, then with my rifles and pistol in the last few years.

When the wrestling season was over, I was feeling pretty good about myself until here came Moses, like right on schedule to make fun of me and get everyone else to laugh at me, too. This day a cadet was found reading a *Playboy* magazine in class behind his textbook. Moses spotted this and took the magazine away from him. Then Moses showed everyone what the cadet had been looking at: a picture of a nude girl with huge breasts.

Moses then went around the whole classroom asking each cadet what this picture represented for him. Many of us were so embarrassed that we didn't want to look at the girl or answer his question, but he kept insisting, and so some cadets finally said sex, while others said Marilyn Monroe. When he came to me, I just shook my head, saying nothing. But he got so mad at me that he finally shoved the magazine in my face.

"What does this remind you of?" he demanded of me, pushing the nude picture into my eyes. "Or are you one of those others?"

"Other what?"

"Others," he'd said, rolling his eyes to the rest of the class. "Come on, speak up!"

"She reminds me of a cow who's had a calf," I finally said, seeing her huge ballooned breasts. "Or a mother dog or pig who's just had a litter."

The explosion of laughter that broke out from the whole classroom was devastating, and the loudest laugh of all came from Moses, himself.

"He isn't even one of those others. He prefers pigs and dogs!"

The laughter never stopped. Moses had just given my classmates an open license to ridicule me. And when they talked about Marilyn Monroe, I had no idea who they were talking about, and so they said that maybe I was one of those others.

I got up and walked out of class. This was it. I was going to go home, get my gun, and come back and kill Moses. Moses shouted at me, telling me that I couldn't walk out of class. I laughed. Couldn't he get it? I'd already walked out of class, and now I was running down to the beach so I could run along the shore all the way to Cassidy Street, up the bluff, and get home.

I was in excellent shape, so I knew that no one could catch me once I had a good head start. A lot of guys were faster than me at the Academy, like Mick McLeans, who was on our wrestling team, but only a few had my endurance. I was going to go home to change my clothes, get my gun, put it in one of my schoolbags, then come back to the Academy, find Moses, and walk right up to him and say, "Captain Moses, I need to show you something." Then I'd drop my bag, whip out my pistol, and gut-shoot him a couple of times. Then when he was down and bleeding, I'd explain to him why he and his type of bully mentality had to die!

And no one would interfere with me because of the loaded gun in my hand. Also, a part of me didn't give a shit anymore if I was caught or not. I could now see that my objective wasn't just to kill Moses and all these other teachers who had abused us, but for everyone in all the whole world to know why.

This wasn't going to be a surprise attack. This had to be a cold, premeditated act, completely well planned, just as it had been premeditated and well thought out to torture and beat us Mexican kids, starting in kindergarten, so we'd be a people, a *gente,* with our heads bowed down to authority forever, thinking we were inferior and worthless. I now realized that this was how you enslaved a people. You didn't just bring them over in chains from Africa. No, you convinced them that they were inferior, not evolved, subhuman, and then when you took off their shackles, so they could go to work, you'd still have them enslaved and shackled inside of their minds for hundreds of years. And this system of teaching was fine with most Anglo teachers, because in the act of convincing us, *los Mexicanos* and the Blacks, we were subhuman, they'd also convinced themselves that they were superior!

Getting home, I was pouring with sweat, having run the whole way, which was about four miles. I tore off my uniform, got into my Levi's, boots, and Western shirt. I put on my Western hat, got my pistol and holster, and went up to the stables to do a little last-minute fast drawing and shooting. And man, was I ever lightning-fast and accurate, too!

Then when I felt ready, I decided to go back home and get my hunting knife, which was razor-sharp. This way I could cut Moses's guts out, after I gut-shot him so that no hospital could save his life. I wanted to make sure that he died a slow, painful death. After that I'd drive over—before the cops came—and do in the playground teacher, and on her, I'd piss in her face as she squirmed in death, and say, "No English, *cabrona!* Spanish only!" Then, if I still hadn't gotten caught, I'd drive out to the San Luis Rey Mission and get that priest who'd tortured me.

I was now sixteen years old, in perfect shape, and had arms of steel. I could do sets of one hundred push-ups with both arms, or sets of forty-five with either arm. And I was an experienced killer. For the last two hunting seasons I'd gone with my father and the Thills to Meeker, Colorado, to hunt mule deer. This last year I'd been so good with my new 30/06 Model 70 Winchester that I'd filled out the tags for almost everyone in our camp, because what most of these he-men hunters really liked to do was drink, play cards, and get away from their wives.

Six huge bucks, I'd gotten this last year. I was a crack-shot with rifle or pistol, and right now I was going to use a little .22 pistol to do in Moses. I'd use hollow-point bullets and split their points in half with my knife for extra quick expansion in gut-shooting him at close range.

I was at our big gun cabinet, getting my hunting knife and extra ammo, when I heard my mother and dad in a big argument. This kind of screaming I hadn't heard for quite a while. I put my pistol, rifle, and knife down, and went to see what the commotion was. My two sisters were in the bedroom with my parents, listening to the argument.

"What's going on?" I asked Linda.

"Dr. Hoskins is dying and *papa* wants *mama* to go with him to see him."

"But why?" I asked. My heart started beat, beat, beating. "He's the reason that our brother Joseph died," I said.

My sister shrugged. "Yes, I know."

"Lupe, please," our father was saying, "it's been years since *Chavaboy's* death, and life goes on. We all need to know how to let go."

"MAYBE YOU! But not for me! I still feel it cutting here inside of my heart! Joseph was only sixteen! He had his whole life ahead of him, when this no-good drunk doctor—"

"Lupe, Lupe, please," said our father. "The man is dying now. And that article in the paper must have cut his heart in two. We need to find compassion in our hearts and forgive him, Lupe."

"NO! NEVER!" screamed our mother. "I hope he BURNS IN HELL!"

The day before, our local newspaper had reported that Dr. Hoskins was caught drunk on the job at the hospital, and talk had it that dozens of people had died over the years because of his gross incompetence.

"Lupe," said our father, "this isn't just about forgiving him. It's also a way for us to bring all this anger out of our own bodies that we've been carrying in our hearts all these years."

"But how dare you say that to me, Salvador! All these years I've carried nothing but love in my heart for our dead son!"

"Yes, but what have you carried for the living, eh?" said my father, then he added, "Tell me, Lupe, how can we expect God to forgive us, if we can't forgive others."

"BUT I'VE NEVER DONE SUCH *SEMEJANTE PECADO* like this monster did to our son in all of my life! Maybe you, who've done so much wrong in your life, can find compassion, BUT NOT ME!" she screamed.

I was shocked. This was entirely not like my mother, who was usually so compassionate towards all people, and soft-spoken, too. But I was also in total agreement with her.

Our father closed his eyes in concentration. "Lupe," he said in a very slow, gentle voice, "Jesus, who was free of sin, they say, He still found it in His *corazón* to forgive those who crucified Him. To forgive, Lupe, isn't really for the other person. It is for helping us find peace in our own hearts. I'm going up the hill to see Dr. Hoskins, and you can come with me, if you'd like."

"NO, SALVADOR!" screamed our mother. "I forbid you to do this! This is CRIMINAL! He's going to get his JUST DESERVE AT LAST! Let him now suffer as I've suffered all these years! MAY HE BURN IN HELL, is what I say!"

I wholeheartedly agreed our my mother. Let the damn cards fall where they fell! Let the son of a bitch suffer just like I was going to make Moses suffer when I now drove back to school and called him out. But my father obviously saw the whole situation very differently, and he now picked up his Stetson, and without saying another word, walked out the door.

My sister Linda ran after him. She'd been following our dad around and trying to act like him for as long as I could remember. When she'd been three or four, she'd ride her tricycle around the front fountain until our father would fall asleep. Then she'd take his cigar and smoke the gigantic thing just like he did as she rode around and around the fountain, talking to the goldfish.

I stayed behind with my mother and our little sister Teresita.

"Damn your father!" said our mother, crying. "Why does he have to talk to me like that, trying to make me look like a bad person in front of you kids! Hoskins is finally getting his just deserves! Don't you see that, *mijo?*" she said to me. "I'm not a bad person. It's just that I can't forgive him in my heart, that would be false!"

"Yes," I said. "I can see that, *mama,* but . . . well, maybe *papa* is right. We all need to learn how to let go. Joseph himself, I'm sure he wouldn't be happy to see you still mourning him after all these years. In fact, the last time I was with him, he told me that what was really needed here in the world wasn't more—"

"YOU JUST DON'T KNOW WHAT IT IS TO BE A MOTHER!" my mother screamed at me, full of anger!

Hearing this, I turned to go. This, I'd heard too many times over the years.

"Where are you going?" she asked. I kept walking. "I'm talking to you!" she shouted.

I stopped and took a big breath. *"Mama,"* I said, turning back around to face her, "I am not going to listen to you tell me one more time that I don't know what something feels like because I'm not a woman or I'm not a mother." My heart was pounding. "That's bully tactics. If you really think I can't feel that, then don't bring it up to me. Hell, sometimes I think that you just like all this suffering, so that then you don't have to go on with your own—"

I stopped my words. Could this also be what I was doing—using all my rage to avoid going on with my own life? I took a deep breath, said nothing more, and turned to go once again.

"Where are you going?"

"I don't know," I said. "Maybe to go with *papa.*"

"MIJITO, PLEASE, DON'T DO THIS! I BEG YOU!"

"Mama, I love you," I said. "But I also kind of, well, got to see what

papa is going to do with my own eyes, because, well, maybe even I, myself, would be a lot better off to let go of all the anger I carry, *mama*."

I was shocked at the clarity of the words that had just come out of my mouth. Hell, I'd never had these thoughts. Where had these words come from, and with such utter clarity? This was when I realized that that little quiet humming had started behind my left ear. It had been a long time since this had happened.

I went out of the house, leaving my little sister Teresita with my mother. I felt sorry for Teresita. In no time at all, I was sure that my mother would have her crying and praying. I got my guns and knife, a bag of ammo, and went outside and got in my brand-new turquoise-colored Chevy pickup, that my dad had gotten for me for my sixteenth birthday, and started down our long driveway.

Getting to California Street, I stopped. I didn't know which way to go. Should I drive up the hill to see what my father was going to do, or should I turn right, go down California Street, then take Hill Street south to the Academy and kill Moses?

I finally put my foot to the gas pedal of my Chevy, deciding to turn left, even though . . . it was way easier for me to go kill Moses. Instead, I went up the hill to Dr. Hoskins's place. Moses would still be at the Academy, I figured, after I was through seeing what my dad was up to, then I'd kill him.

Getting to Dr. Hoskins's place, I saw our big car parked to the left of his house over by the riding ring. Dr. Hoskins didn't ride Western. He rode English, or jumping, or some such stuff. I parked my turquoise-colored Chevy alongside my dad's navy-blue Cadillac and got out. And there was my father and my sister Linda talking to Dr. Hoskins inside of the riding ring. They were talking quietly as people rode around them on horseback. I could hear that they were talking about saddles and bridles and the best way to keep the leather soft and clean. They were visiting, like old friends, and I could see that the doctor looked so grateful that our dad had come to see him.

They weren't saying a single word about my brother. And they weren't talking about the newspaper article, either. Then, when I came up close and my dad saw me, he introduced me to this man, whom all of my life I'd always known as the monster.

Dr. Hoskins took off his riding glove and offered me his hand and I could see that he was nothing but a tired, old gentleman who looked very close to death himself.

I hesitated. I glanced at my dad, and he nodded to me, but still, it was real hard for me to reach out and take the doctor's hand. After shaking his hand, I immediately backed away, and watched my dad and this old doctor continue visiting. I flashed on what my brother had told me on the last day we'd been together. He'd told me what was needed on Earth wasn't control or more money or great new inventions. No, what the world really needed was so simple—patience, love, compassion, forgiveness, and understanding.

Remembering this, I looked at my dad and I could see that this was exactly what he was doing. And he wasn't faking it, either. *Mi papa* was really, really forgiving this man inside of his heart as he spoke to him. I just couldn't, for the life of me, figure out how my dad was able to do this. Hell, I was ready to drive over to school as soon as we were through and kill Moses.

I wanted REVENGE! I wanted JUSTICE after all these years, just like my mother, I WANTED MOSES TO BURN IN HELL FOR ETERNITY!

And yet, as I stood here and watched *mi papa,* a child of the Mexican Revolution, a man who'd gone to hell and back so many times in his life that he usually had more rage and vengeance in his heart than any ten *hombres,* I could see that he was setting aside all this as he now spoke with the doctor.

Then I saw it so clearly, why my dad was "forgiving" this man, just like Jesus "forgave" those men on the cross, because they hadn't known what they'd done.

My heart began beating! No, no, no, I could never, never, never do this, AND LIVE! I had to kill Moses and all those other people before I could live.

But then it hit me. How in the world did I expect to live after I killed

Moses and all those other people? The cops would come to get me and no, I wouldn't kill them because I had no beef with the Oceanside Police—in fact, three cops were close friends of ours—and I'd be arrested and taken to jail. And my parents would lose another son. Then, both of their sons would be gone.

So then, maybe killing wasn't the answer. Maybe there was another way to get the world to know about the abuses that had happened to me and were still happening to so many *Mexicanos?*

My heart was beat, beat, BEATING! "Well, then, could it be," I said to myself, "that my father was right when he'd told my mother 'yes, you've been carrying love for Joseph, but what love have you been carrying for the living?' " Because it was true, I could now see that our mother had been carrying her love mostly for the dead ever since my brother had died. And me? What was I doing?

I breathed, and I breathed again and continued to give witness to my dad, to *mi papa,* as he listened to all this doctor's small talk about bridles and saddles with such kindness, patience, compassion, and yes, forgiveness. And I could now see that this didn't make my father look weak. No, it was the opposite. It made him look so strong and healthy, twenty years younger than this doctor, and yet, I bet that they were pretty close to the same age.

My eyes started watering. I could now also see that this doctor was so grateful for this opportunity of "forgiveness" that it looked like he was going to leap out of his skin and hug and kiss our dad for having come to see him. I took another breath and wondered if I'd ever be able to do what my dad was doing, or would I forever be so angry, here in my heart, that I couldn't forgive just like my mother couldn't.

I turned. I'd had enough. I said goodbye and went to my truck. I needed to go to the Academy real quick and gut-shoot Moses before I lost my nerve!

After all, I was tough! I was a wrestler! I'd killed lots of game! I was no little scared coward anymore! But then, I remembered that whatever I'd been through, my dad, *mi papa,* had been through a thousand times more, and yet he was finding it in his heart to forgive this monster who'd killed his son.

"Joseph," I now said as I drove down California Street and turned left on Hill to go over to Carlsbad to kill Moses, "help me. I don't know what to do, and it had all been so clear to me before I saw our dad talking with Dr. Hoskins with such peace in his heart."

Instantly, I felt that old purring behind my left ear. I'd asked for help and here it was, humming, purring behind my left ear, and then traveling to my right ear. Then I heard that little Voice inside of me and it was telling me so clearly—without any confusion whatsoever—to turn right on Cassidy Street and drive down to the beach.

I had no idea why I was doing this, but I trusted this little inner Voice of mine, so I did as I was told and turned right on Cassidy. I crossed over the railroad tracks, heading for the beach. Then I was told to turn right on Pacific and go past Buccaneer Beach. I did as told. Then at the top of the first little hill, just past Buccaneer, I knew that I was supposed to park and get out of my Chevy. I did, leaving my guns in the truck, crossed the street, and walked out on the bluff. And there before me, sticking out of the incoming surf, was that big black rock.

I hadn't been down here in years. It was the lowest tide I'd ever seen. The rock was more visible than ever. Quickly, I ran down the bluff to the water's edge. Back then there were only a few houses in this area. All of the beach houses were either south or north of here. I glanced around, saw no one, quickly stripped, and walked out into the surf. The water was cold and felt good.

I walked out a ways, then dove under the first good-sized wave. I'd become a pretty good swimmer in the last couple of years. The water was delicious! I swam out to the big rock, but didn't get too close, so that the waves wouldn't smash me up against it.

This was when I first saw the fins. And they were coming right towards me. The dolphins began weee-heeing to me just like they'd done to Midnight Duke that day years ago. I rolled low guttural sounds back to them and they began chirping.

I dissolved. Just like that, all my rage dissolved inside of me, and I started swimming out to them. They screeched with happiness. I screeched, too. This was all a dream come true. They came close and began to play with me. I laughed, I was so happy. Big, big, BIG HAPPY! We began to talk together like *familia*. To give Song back and forth. I, too, was now helping *Papito* paint His Garden of *Paraíso!* I'd found my place. I was free.

Afterword

About my reading *problema;* I didn't find out that I was dyslexic until about 1985 when my wife and I took our two boys to a reading special-ist because they, too, were having reading difficulties. One of our sons tested slightly dyslexic, the other tested moderately. The test scoring went from 1 to 20, with 20 being the most severe. Both of our sons were in the 8-to-12 range. We were told that dyslexia was a catchall phrase, and that some forms of dyslexia were hereditary. I decided to get tested, too, but I figured that it wouldn't be a true test for me any-more because by now, at forty-five, I knew how to read pretty well.

I took the tests. When the woman practitioner returned with my re-sults, I could see that she was on the verge of crying. She told me that I was completely off her charts. That it was a miracle that I'd ever learned to read, write, or even listen because I had both visual and audio dyslexia. I began to cry, too. Someone finally understood all the "hell" that I'd been through since a child when I'd first tried to understand language. And yet in other forms of communications, like painting, sculpture, music, math, problem-solving, and chess, I'd been very good. In fact, in high school, once I learned how to play chess, I'd play lightning-fast, intuitively seeing all these different possibilities at the same time, and I'd won well over a hundred chess games without los-ing a single game. And that included beating some of our faculty mem-bers who thought that they were very good at chess.

So then, what does this catchall phrase dyslexia really mean? Is this what enabled me to feel that humming behind my ears? Was this what allowed me to sometimes see the whole world come alive in light and color? Could dyslexia be a gift? Could it be that we were all "dyslexic" back at one time when we all recognized that the Kingdom of God was within and we knew how to bring what was within out into the world?

Then about three weaks ago, now in the year 2003, I received an e-mail that I think has maybe given me a small part of the answer.

> Aoccdring to rscheearch at an Elingsh uniervtisy, it deosn't mttaer in what order the ltteers in a word are, the only iprmoetnt thing is that the frist and lsat ltteer are in the rghit pclae. The rset can be a total mses and you can still raed it wouthit a porbelm. This is bcuseae we do not raed ervey lteter by it slef but the word as a wlohe and the biran fguiers it out aynawy.

Could it be we stifle our children's genius by languaging them too quickly away from their hearts and into the straight and narrow confines of linear thinking? For thousands of years women have been told that they become too emotional if they get into their feelings. And men have been told that they are wimps if they get into theirs. And yet could it be that only by getting out of our "heads" and into our "hearts" and "souls" can we access our Intuitive Genius—our Spirit Guide reconnecting us to our natural full Thirteen Sensory Perception?

About "English only": Let me tell you a little story that happened in Houston, Texas, just before I became a national bestselling author. I met this young woman at the University of Houston. She looked like she was part Black, part White, and part American Indian. She was stunningly beautiful, with huge greenish eyes. She spoke Spanish. I asked her where she was from. She said Panama. I asked how she liked the United States. She said she didn't, and that as soon as she graduated she wanted to return to Panama. I asked why. She told me that she'd had a boyfriend for four years. "And the other day he said, 'I think I love you,' so I dropped him as fast as I could. My God," she added, "after four years he was still thinking about our love. I can't stand to be around people who are always thinking so much."

I laughed. I could see her point completely, because in Spanish you'd never say, "I think I love you," especially after four years. That would be an insult. You'd say, "I feel love for you so deeply that when I just think of you, I start to tremble and feel my heart flutter." Why? Because Spanish is a feeling-based language that comes first from the heart, just as English is a thinking-based language that comes first from the head. And Yaqui, Navajo, and the fifty-seven dialects of Oaxacan are ever-changing languagings that come first from the soul, then go to the heart, and lastly to the brain.

In one dialect of the Mayan languaging down in the southern tip of Mexico in the Yucatán, I was told that there are twenty-six ways to say "love," just as in the Eskimo languaging—I've been told—there are twenty-six ways to say "snow." And so when a married man or woman has an affair with some other person there's actually a word for this kind of short, little, happy love, just as there is another word for the kind of long, harmonious, thorough love that brings the married couple back together without having feelings of being betrayed. In fact, betrayal in marriage isn't a concept that's even available within this Mayan dialect, because love, in all her great wild twists and turns, is forever growing, changing, deepening in the on-going drumbeat of our eternal hearts and souls.

So could it be that we live in a very small and limited world by only speaking one language? Could it be that "only one" of anything imprisons the mind—religiously, socially, and politically—and only one in language is the first sign of the end of any nation?

And also, could it be that in the first part of their lives, kids can easily learn two and three languages at the same time, but orally, and with play and games and healing stories; thus expanding their mental capacity far beyond what we now call the norm, and access genius. Could it be that being taught to read and write before children have had their fullness of dream and play and adventure stifles them for life? Could it be that the whole future of our country, here in the US of A, is dependent on how fast we can get out of the straight and narrow confines of

"English only"—which had its place for a little while—and enter into a greater and more flexible global understanding of communications?

Personally, I had to go six months without speaking as an adult before I was able to re-find my Inner Voice to become a writer. And in that time I came to many very interesting understandings, one being that maybe, just maybe, even Jesus, Himself, deliberately didn't learn to read or write until he was well past twelve years old and that everything that He knew of the Bible before that age had been read to Him? Why? Because He hadn't wanted to clutter his mind and get His heart and soul bogged down with the details of linear thinking if he was going to accomplish His Earthly Work in such a short life. Believe-you-me, children truly are our latest messengers from God, as my grandmother used to tell me, and each is unique and wonderfully brilliant already.

One more story, about two months after my talk in Long Beach, California, at that CATE convention, I saw Moses coming out of the Carlsbad post office. Immediately, I tried to walk around him, so he wouldn't see me, but he turned and saw me.

"Hey!" he said. "Aren't you the—the author?"

A lot had been written about me after the CATE conference in most of the Southern California papers, plus the Bay Area of San Francisco.

"Yes," I said. "I'm the writer."

He smiled and looked at me, like he was trying to place me. He looked old and weak, and certainly not as big as I remembered him.

"I had you in class, didn't I?" he asked.

I nodded. "Yes, you did, sir."

"I thought so," he said. Then he said something that I'll never forget as long as I live. "We had fun, didn't we?" he asked.

Hearing this, I almost shit a brick. I guess he had no memory of the torture that he'd put me through. I smiled. I laughed. Then did this mean that I was the only one who remembered all the crap that he'd done to me. I felt like belting him in the mouth. But how could I do this, he looked so old and weak and pitiful, just the way Brookheart had referred to him.

"Yes," I finally said, "we had fun, sir."

"I thought so," he said, grinning.

I began to grin, too. What else could I do? "Your wife," I said, "how is she?"

His face twisted. "She died several years back," he said.

"Oh, I didn't know," I said. "I always liked her. She was a very decent, goodhearted woman."

His old wrinkled-up eyes filled with tears. "She was the best thing that ever happened to me."

I nodded. "I bet you're right. 'Bye," I added.

"Goodbye," he said. "Good to see you!" he shouted after me. "Keep up the good work! We're all proud of you!"

When I was done at the post office, I crossed the street, went into the restaurant bar across the street, and ordered a beer.

"To You, God," I said, making a toast, "to You and Your wild sense of humor." I drank my beer, then got on my bike and pedaled over to the beach. A part of me realized that I had to thank Moses, because it was a lot of my hate towards him that had kept me going all these years.

Down at the beach, I pulled over and looked out at the sea. Past the surfers, the water was calm. There really weren't any good waves; not for surfing, anyway. I breathed and became mesmerized by the quiet movement of the water as I looked out past the breakers to the huge expanse of ocean, our mother, the place from where all life originally came.

I said a little prayer for Moses and his wife. Then it dawned on me that his name was just like Moses of the Bible. I'd never made the connection before, and it was so obvious. I started laughing. Oh, my Lord God, there really was a larger, grander plan going on that we mortals couldn't see. Moses had done his job once again. It had been "he" who'd led me out of the desert and to the promised land within my own soul.

That little purring began behind my left ear, traveling across the base of my head to my right ear. All around me, I began to see the whole world come to life with light and color. Once again it was another gorgeous day in *Paraíso*.

Thank you, *gracias*.

Acknowledgments

First of all, I'd like to acknowledge you readers who have been purchasing and reading my books all these years. Without you I would have no career and there wouldn't be any books. Thank you, *gracias* from *mi familia* to all of you and your *familias*. And a special thanks to you readers who have written to us or e-mailed us, because, even though we haven't answered all of you, we have read and treasured your letters. One reader from Washington, D.C., told us that he read *Rain of Gold* to his father every afternoon in the hospital for the last two weeks of his father's life. He told me that the book bonded him and his father across generations as no other experience ever had. Thank you. I am humbled, as we are told story after story of how these books of our *familia* have taken on a life of their own, validating the stories of our readers' *familias*. Truly understand, it is YOU WHO HAVE VALIDATED the heartfelt struggles of our crazy-*loco* family! I salute you, *con todo mi corazón!*

Also, I'd like to thank my longest living editor Rene Alegria; well, it's not that he's really that old, it's that he's been the only editor who's "lived" this long with me. I love you, Rene, and respect you, and yes, hate you at times, too, but that's what happens when you become *familia*, just as my longtime agent *Margarita* McBride is *familia*, seeing us through thick and thin and even thinner. Thank you, Margret, and your great staff, Donna, Renee, Faye, and Anne. And thank you, Andrea *de Colombia*, Rene's assistant editor. And also a big thanks to Gary Cosay, my movie agent, and Chuck Scott, my longtime lawyer, two people who have been with me for over thirty years, and my new lawyer Mark Hollaran.

Then I'd like to thank my sister Linda, who's now been running my office for speaking engagements for several years. After all those years of brother-and-sister arguments, we're doing pretty damn—I mean blessed—well. And thanks Jackie, who handles my finances, and Jolyn, who's been typing for me till midnight, and last but not least, thank you Juanita, *mi esposa,* for keeping calm even when I jump up at two in the morning and I'm so full of the Great Spirit of writing that I can't keep still. Thanks, Juanita.

And now I thank God, *Papito,* and my brother Joseph, my dad and mom who've passed on, too, and my Spirit Guides who were assigned to me at birth. *Gracias, mi Angelitos de Luz,* and our first grandchild, Isaac Salvador, meaning "Laughing Savior."

Victor E. Villaseñor
PRESENTATIONS, SEMINARS, AND WORKSHOPS

Since Victor Villaseñor's first public presentation at the CATE conference in Long Beach, California, in the early 1970s, he has been sharing with audiences across America his passion for life and vision for world harmony. His topics of family, "herstory" instead of "history," and global peace are told through personal stories filled with emotion, humor, and total abandon.

His own personal story—childhood, the *barrio,* schools, and Mexico—formulates his unique message. Through his own experiences, he passes on the knowledge for "creating a brave new self with honor and respect." He welcomes you into his ancestral vision of one race, the human race, and that we were all an indigenous people at one time—connected directly to Creation through our natural Thirteen Senses.

Victor Villaseñor's personalized presentations are especially designed for parents, teachers, students, community and business groups, and for anyone interested in learning how to make lightning-fast, intuitive decisions of true genius. With his gift for storytelling, Victor Villaseñor empowers his audiences with hope and self-worth, helping them reach new heights of trust, vision, and cooperation—a feeling that we are all supported by the universe, and that we are one united people. Standing ovations are standard, following Victor Villaseñor's talks. For more information about Victor Villaseñor's presentations, seminars, and workshops, please go to his Web site at www.victorvillaseñor.com. To contact Victor Villaseñor, e-mail him at victor@victorvillaseñor.com, phone him at 760-722-1463, or fax at 760-439-1204.